M

SCIENCE

GEOLOGY

HarperCollins books may be purchased for educational, business, or sales promotional use. For information, please write: Special Markets Department, HarperCollins Publishers, 10 East 53rd Street, New York, NY 10022.

Produced for HarperCollins by:

Hydra Publishing
129 Main Street
Irvington, NY 10533
www.hylaspublishing.com

FIRST EDITION

The name of the "Smithsonian," "Smithsonian Institution," and the sunburst logo are registered trademarks of the Smithsonian Institution.

Library of Congress Cataloging-in-Publication Data

McMenamin, Mark A. S.
  Science 101 : geology / Mark McMenamin. – 1st ed.
      p. cm.
  Includes index.
  ISBN: 978-0-06-089136-7
  ISBN-10: 0-06-089136-X
  1. Geology–Popular works. I. Title.

  QE31.M394 2007
  550–dc22

                              2007060875

07 08 09 10   QW   10 9 8 7 6 5 4 3 2 1

SCIENCE 101

GEOLOGY

Mark A. S. McMenamin

Collins

An Imprint of HarperCollinsPublishers

CONTENTS

Agate

Quartz

WELCOME TO GEOLOGY

Left: Both the formation of its travertine terraces and the thermal history of Mammoth Hot Springs in Yellowstone, Wyoming, are fertile ground for geological study. Top: The science of geology examines the particulars of a single rock, such as this grossular garnet, as well as whole terranes. Bottom: Mount St. Helens in Washington State. Volcanology is the branch of geology that studies volcanos.

Geology is the science of the Earth. It is also the science of rocks, and the science of tiny sediment particles. It is the science of the atmospheric gases released from Earth's interior, and also the science of the minerals that absorb those gases as the minerals begin to weather away. It is the science of ancient life-forms and their past environments, and is also the science most closely associated with the search for the origin of life. It is also rapidly becoming our planetary insurance policy as we scramble to decode the secrets of climate change and agonize over ways to deflect killer asteroids before they slam into the planet.

Finally, geology is the science of wonder. Imagine that you have just finished this book and you are walking along the shore of a lake. You pick up a brightly colored cobble that has been smoothed and rounded by abrasive action of stream transport. Dusting the sand off for a closer look, you note patterns and texture. Variant hues in the rock are due to metamorphic foliation, wavy bands that mark out distinct, shimmering layers. With your knowledge of geology, you realize that the rock you are holding is not merely a pleasing souvenir but rather a glittering archive of Earth history.

The details of this particular rock's history may be unknown to you, but you can read the general outline of its history as plain as this page. The rock is now composed of metamorphic minerals, minerals that were wholly or in part transformed from the crystals in the original rock that had been metamorphosed. The rock you hold, therefore, has had two incarnations, first as the rock prior to metamorphism (protolith) and second as a metamorphic rock. The separation in time between the two could be as much as 100 million years or even more. The rock may in fact show multiple phases of metamorphism, but this will be difficult to determine by direct visual inspection. One thing is certain, however: the cobble is in the process of undergoing a third transformation. This is mechanical and chemical breakdown of its surface and interior, a process that began when it separated from its parental bedrock. The separation launched a rock as an angular, larger chunk that tumbled down a slope. The rock was carried by gravity and water down a series of inclines until it reached a streambed. There, the rigors of stream flow bounced the rock against many others of similar size and shape; all were rounded and transported tens of miles or more away from their points of origin. The smooth rounding of the rock is only the latest in a series of stages that, in aggregate, have molded the rock and transported it to you. You hold in your hand a fragment of Deep Time. In your mind's eye you can roll the cobble back upstream, restore it to its original outcropping, push the rock ledge back into the Earth whence it came, unmetamorphose the minerals back to their pristine conditions in the protolith, strip the sedimentary cover layers off the protolith layer, dissolve the cement that binds the sediment particles together, scatter the sediment particles backward in time and, if you have not yet grown "giddy at the prospect" as John Playfair (1748–1819) put it, follow each one of those tiny sediment particles back to its parent rock, and then restore each one of these parent rocks back to its living outcrop and in doing so restore the surface of Earth to that of a bygone age. As this is indeed getting to be a bit much to contemplate, you grasp the edges of the cobble, cock your wrist, and fling the stone across the lake surface. It is a good toss, resulting in a satisfying seven skips and seven sets of converging, concentric wave fronts. Thus, the rock cycle stretches back over billions of years in a wonderful fashion.

MODERN GEOLOGY

Modern geology still grapples with the immensity of geological time. There is a lot of it. Geologists take it for granted. To most geologists one million years is not a great deal of time, and to Precambrian geologists even ten million years is merely an instant. Geologists are nevertheless obsessed with time and its measurement. The minerals of your stream cobble hold clues that can allow the timing of metamorphism to be determined with a certain degree of precision. With any luck, it would be possible to determine when the protolith was deposited (assuming it was sedimentary) and what kind of rock it was. In some cases it might even be possible to identify fossils in the rock, such as burrows and trails that, as part of the fabric of the rock, were able to withstand the transformations of metamorphism. A good forensic scientist works much like a geologist, leaving no stone unturned in the search for clues.

Tufa towers, Mono Lake, California.

Norway's Briksdale Glacier.

Science 101: Geology spans a spectrum of scales in time and space, from processes that occur over spans of billions of years to events that happen in a few seconds, from galactic clouds to fissioning atomic nuclei. This book also depicts the influence of life on Earth, and provides ample evidence for geochemist Vladimir Vernadsky's view that life makes geology on Earth. Chapter 1, "Earth the Planet," describes the origin of the planet and its primordial interior heat sources. Chapter 2, "Planetary Geology," provides a lesson in comparative planetology that sheds important light on Earth itself. Chapter 3, "Minerals, Rocks, and the Crust," will guide you through the essentials of silicate minerals; these mineral types and the rocks they form are the backbone of planetary geochemistry. Chapter 4, "Weathering and Soils," charts the breakdown of these very same silicate minerals in a set of processes that are unique to Earth and also uniquely important. Chapter 5, "A Living Planet," demonstrates that whereas silicate minerals may form Earth's skeleton, it is microbes and their biogeochemistry that really flesh out the planet. Chapter 6, "The Fossil Record," traces the evolutionary development of life from microbes to complex life in all its iterations. Chapter 7, "Go with the Flow," sketches out the hydrologic cycle and its influence on the rocks of Earth's crust. Chapter 8,

"Glaciations through Time," examines the intervals when much of the hydrologic cycle is taken out of circulation, so to speak, and the Earth takes a turn for a colder, and in some cases, very much colder climate. Chapter 9, "Geological Catastrophes," identifies short-lived phenomena that carry with them potentially devastating consequences. Chapter 10, "A History of Geological Thought," examines a series of geological controversies that illuminate how we know what we know about Earth. Chapter 11, "The Plate Tectonic Revolution," charts the transition from continental drift theory to plate tectonic theory, in what must now be viewed as a great revolution in scientific thought. And last, in Chapter 12, "Geology in the Field," you will find an introduction to both classical and newly developed techniques in this fieldwork-intensive science. These chapters are supported by an extensive Ready Reference section and a glossary of terms. With recent advances in biogeochemistry, paleontology, mapping, and planetary geology, we have reached a unique time where

You hold in your hand a fragment of Deep Time.

the structure and history of our planet can be seen with a richness that was not possible until now. We cannot follow them all, of course, but we are now able to trace the most fascinating trajectories of the skipping stones of our planet.

CHAPTER I

EARTH
THE PLANET

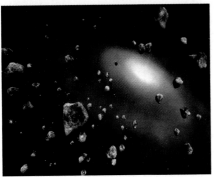

Left: Artist's conception of planet formation. It is thought that the planets formed from a rotating disk of material that formed around and at the same time as the Sun (center), around 4.6 billion years ago. The rocky inner planets formed during a chaotic period when many protoplanets formed by gathering up surrounding material (such as the rocky debris seen) by gravitational attraction. Eventually, four survived, along with numerous asteroids. Top: Nebula with stars. Bottom: A rendering of deep space, with asteroids and potential comet nuclei in orbit around a young sun.

Planet Earth is part of the solar system, and this solar system is literally formed of star dust. Cataclysmic explosions of early-generation stars distributed the heavy elements that were created by fusion inside the star. As these materials were strewn into the universe they gradually began to congregate once again, thanks to gravitational attraction. Along with less-dense matter, such as hydrogen and helium, heavy elements helped to form the rotating cloud of gas and dust called the presolar nebula. Heavy elements such as iron and nickel tended to stick together, and in sizes ranging from dust particles to moon-sized asteroids these refractory materials found themselves orbiting the massive center of the nebula—that is, if they escaped falling into its center. The rocky planets that orbit our Sun represent local gravity wells: in other words, accumulations of solid mass that attract other massive solids. Earth, then, seems to have grown bit by bit in a process that continues to the present each time a meteor or speck of cosmic dust enters the atmosphere. In the early days, however, there was nothing gradual about the collisions of meteors that must have fused and melted the earliest rotating glob of material that we could have identified as Earth.

Planetesimal Impacts

In "The Method of Multiple Working Hypotheses," a charming and now classic article published in 1890, geologist Thomas Chrowder Chamberlin (1843–1928) argues that scientists should keep a family of hypotheses in mind rather than fixating prematurely on a preferred explanation in any case of scientific uncertainty. This admonition did not prevent Chamberlin himself from eventually arriving at a preferred hypothesis of his own, along with American astronomer F. R. Moulton (1872–1952): the Chamberlin-Moulton hypothesis for the planetesimal origin of Earth. Chamberlin's method of multiple working hypotheses seems to have served him well, for the Chamberlin-Moulton hypothesis is now

Below: The surface of the Moon (lunar regolith), showing Moon rocks. Top left: Artist's conception of a bolide (large asteroid, meteor, or comet) collision on early Earth.

Geologist T. C. Chamberlin, remembered today both for his writings on the scientific method and for his early articulation of the planetesimal hypothesis.

widely accepted as the best explanation for the primary process that formed our planet. There remains, however, some uncertainty about how this planetesimal formation could have occurred. Two major questions are associated with the planetesimal accretion theory: First, how did the planetesimals form in the first place? Second, once planetesimals grew large and began to collide, what prevented the great amounts of energy released on impact from blowing everything to bits? Actually, these questions have closely related answers.

MOON ROCKS

At the first *Apollo 11* Lunar Science Conference in January 1970, S. K. Asunmaa, S. S. Liang, and G. Arrhenius, fresh from their studies of the newly arrived lunar rocks provided by NASA, grappled with the first question: Namely, how did primordial accretion occur to form the planetesimals? Somehow, the primary grains needed to coalesce into large pieces, but the gravitational attractions between the little grains were so small as to be negligible. Electrical charges might bring the particles closer together,

but by themselves these forces were not strong enough to fuse the particles into a solid mass, and, in any case, electrostatic forces are too weak to keep grains together with relative velocities of greater than about one meter per second. Asunmaa, Liang, and Arrhenius concluded that the grains must have been welded together in the period of time after a high-velocity impact when relative velocity was decreasing to the point that the hot grains (with still molten surfaces) could weld together. Using scanning electron microscopy to view particles of lunar soil, they identified a variety of "coatings and bridging structures between individual particles, thereby developing an aggregate structure with increasing cohesion."

THE ACCRETION EVENT

Radiometric dating (using rubidium-strontium and neodymium-samarium dating methods) of the most ancient meteorites suggests that planetesimals were formed more than four and a half billion years ago. Once formed by the welding-accretion process, the planetesimals came together by a process of impact and collision. Such great amounts of energy were released in these impact events that one might expect that the impact explosion would overcome any gravitational or electrostatic attractions and blast everything (both impactor and its target) back into space as a hot dust (where the welding process would begin anew).

Indeed, this does seem to have been at least partly the case during the formation of Earth's Moon. The leading hypothesis suggests that the early Earth was hit by a giant planetesimal approximately the size of Mars. This impact led to the ejection of a huge amount of the early Earth's mantle into space, to form what for a short time at least must have resembled a Saturn-like ring of superheated rock and dust. The welding process led to the formation of the Moon in fairly short order,

giving Earth a natural satellite that was considerably closer to Earth than it is now (it has been moving away ever since). Smaller planetesimals, in contrast, did not have such a dramatic effect on the early Earth itself. Nevertheless, the energy delivered with each impact was immense. But instead of being Earth-shattering, calculations have shown that the energy from these smaller collisions would mostly be transformed into thermal energy that was largely retained within the early Earth.

FORMATION OF THE MOON

The leading theory for the origin of the Moon is called the Giant Impact Hypothesis. Proposed in 1975 by W. K. Hartmann and P. K. Davis, this scenario posits an enormous impact between the early Earth and a Mars-sized planetesimal at a collision speed of around 15 kilometers per second. The material ejected from proto-Earth formed a ringlike structure that, just beyond the Roche limit surrounding Earth, gradually accreted to form the Moon. Much closer to Earth than it is now (it continues to recede with time), the Moon formed a core and mantle as did Earth. Due to its small size, the Moon cooled more rapidly, and insufficient internal heat exists in it at this time to generate much, if any, lunar volcanic activity.

The Earth-Moon system as seen from a near-Earth vantage. Gravitational attraction to the Moon results in ocean tides.

The Four Heat Sources

The heating from planet-esimal impacts was (aside from radioactivity) the first major source of heat present in the early Earth. The larger the planetesimal, the greater amount of heat delivered to the growing planet for two reasons. First, a larger rock body at a given incoming speed brings with it a larger amount of kinetic energy that can be transformed into thermal energy or heat. Second, there will be a difference in the heat retained by the impact of a

into the early Earth, and the incoming energy would be effectively trapped by the surrounding rock of the planet. Some calculations suggest that impacts of this nature could have led to wholesale melting of the early Earth once it had reached a mass of approximately 20 percent of Earth's present-day volume.

METALS AND SILICATES

Planetesimals large and small brought two main types of materials to help grow the early

silicate and metal components in the early Earth is an unstable mixture, and the denser metals would tend to work their way to the core. As soon as the early Earth had sufficient heat from impact to melt, the migration of metals to the core was enhanced as the metals literally began to flow to the center of the Earth.

This flow of metal to the core released tremendous amounts of heat. It is fairly easy to imagine the frictional heat that might be generated as a solid chunk of iron-nickel meteorite slowly works its way past silicate masses toward the core. It is somewhat less easy to visualize this in the liquid state. Yet, the fact remains that if the entire early Earth were to melt, dense matter such as iron would flow to the core and release a vast amount of heat as potential (or gravitational) energy and be transformed into thermal energy. Some calculations suggest that melting and core formation had begun when Earth was only about 20 percent of its present size.

Left: A Martian achondrite meteorite. Right: Chondrite. In some chondrites, patches of iron separate the small grains called chondrules. Different types of chondrite meteorites have different amounts of metal and have been heated to varying degrees. Top left: An asteroid en route to a collision with Earth.

large body in comparison to that of a small body. A rock body such as a small asteroid will deliver most of its energy to the planetary surface, where much of the thermal energy will escape back into space. A large planetesimal, on the other hand, would penetrate

Earth. These materials are silicate rocks, as seen in stony (chondrite and achondrite) meteorites, and metals, as in iron meteorites. Combinations of the two types, incorporating both silicate and metal components, are known as stony-iron meteorites. The combination of

GRAVITATIONAL AND RADIOACTIVE HEATING

As this size and mass increased over time, gravitational compression gained increasing importance as a third heat source, as crushing masses of rock pressed down on one another. From the very beginning, and with steadily increasing intensity over geologic time, radioactive heating also contributed to the planetary heat budget. We know that the intensity of radioactive heating has declined over time, because amounts remaining of, say, the various isotopes of uranium are less than there were earlier in Earth's history.

These radioactive elements are the spent fuel from the nuclear fusion reactions in the centers of stars, and once formed they are not produced again until a new cycle of supernovae renew the process of nucleosynthesis. Thus, some of the isotopes that heated the early Earth are no longer seen to occur naturally today. For example, the radioactive isotope aluminum-26 was unknown on Earth until some was created during the nuclear tests of the 1960s. Indirect evidence within feldspar crystals in chondritic meteorites (in the form of anomalous amounts of magnesium in the extraterrestrial feldspar crystal structure) indicates that aluminum-26 was once present in significant amounts. This would have been true as well in the planetesimals that formed the early Earth, which (even before they impacted Earth) were being warmed internally by the radioactive transformation of aluminum-26 into its daughter product magnesium. With a radioactive half-life of only 730,000 years, the isotope aluminum-26 would have decayed to background levels only about 10 million years after having been formed in a supernova explosion.

Top: Spiral Galaxy M74, 30 million light years away from Earth. Study of this galaxy helped to confirm the hypothesis that supernovae can produce cosmic dust particulates. Bottom: A cutaway view of planet Earth, showing its layers and highlighting the boundary between the core and the mantle.

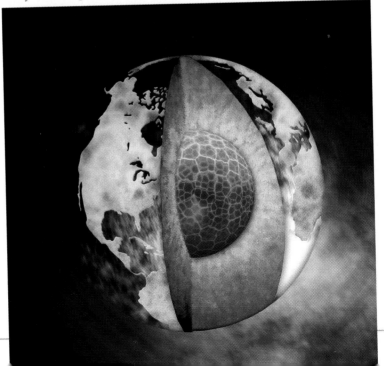

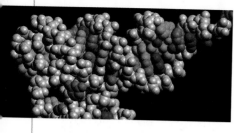

Origin of Life

The problem of the origin of life has long perplexed scientists and philosophers, and the puzzle continues to this day, for no one has discovered the laboratory recipe for transforming simple chemicals into a living cell. Attempts to scientifically address this problem have been made successfully, but progress has been slow. A milestone in the effort to understand the origin of life was made in the early 1950s when Stanley Miller designed the now-famous Miller-Urey experiment, in which simple chemicals such as methane, water vapor, and ammonia gas were exposed to a spark from a discharge tube sealed in a flask. The astonishing thing about this conceptually simple experiment was that it yielded a large quantity of organic molecules, most notably amino acids of a variety of types. Miller and his graduate adviser, Harold Urey, were astonished by the amount of organic matter that formed (it accumulated as a brown gunk on the side of the flask), and this was a great step forward for understanding how nonliving matter could undergo a natural transformation to life. A problem remained, however. The organic soup that was created by the experiment was toxic to life, because it contained amino acids in both the left-handed (*levo-*) and right-handed (*dextro-*) configurations. All known life relies on levo-configuration amino acids, and the dextro version is deadly poison. Although crystal surfaces have been suggested as a possible solution, no high-yield geochemical process has been discovered that can separate the right-handed from the left-handed amino acids.

CYANIDE POLYMERS

A possible solution to the problem was proposed by Clifford Matthews of the University of Massachusetts at Amherst. Matthews noted that the surfaces of comets, asteroids, and other solid bodies in the solar system were covered with organic gunk not unlike the product of the Miller-Urey experiment. Matthews proposes that hydrogen cyanide (HCN) in this mixture has the ability to directly polymerize into proteins, what Matthews refers to as heteropolypeptides. This proposal nicely gets around the toxicity of the two enantiomers (mirror-image) types of amino acids, because each hetero-polypeptide that forms can take

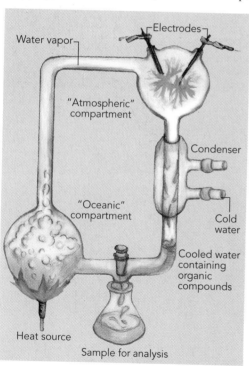

Above: A diagrammatic sketch of the ground-breaking Miller-Urey experiment, depicting the electric discharge in the "atmospheric" compartment that generated chemically produced organic compounds. The yield of organic compounds produced in this experiment is surprisingly high. Top left: A computer-drafted rendering of RNA structure.

only one type of amino acid. Thus, spontaneous protein formation itself sorts out the amino acids. Two types of proteins can form, of course, but it may be a less daunting geochemical task to sort out the proteins (with amino acids already in place) than to sort out the amino acids from scratch.

RNA WORLD

The next step in biopoesis (spontaneous generation of life), of course, is to form nucleic acids and then organize them into the famous chain- or ladder-shaped molecules such as RNA and DNA that can carry the genetic information essential for organized protein synthesis and cell reproduction. This leads directly to a famous chicken-and-egg problem: Which came first, the proteins or the nucleic acids? In a scenario called RNA World, it is the nucleic acids that came first, because experiments have shown that RNA can act as its own self-catalytic molecule. Proteins called enzymes ordinarily do this job, but special bits of RNA called ribozymes can, by themselves, catalyze RNA chains from a solution containing free nucleic acid constituents of RNA.

Still, the task of understanding the origin of life remains a daunting scientific challenge, and recent investigation is focusing on the special geological conditions of the early Earth that might have been particularly conducive to the origin of life.

Colored scanning electron micrograph shows Geobacter metallireducens *bacteria (green) being used to digest uranium waste. The bacterium is able to survive in radioactive environments and turn the uranium waste from a soluble form (that can contaminate water supplies) to insoluble form.*

GEOBACTER AND URANIUM OXIDATION

Many microbes live beneath the Earth's surface in soils, rocks, and sediments that have accumulated on the bottom of lakes and seas. Many of these microbes make a living in the same way that they have for billions of years, namely, by stripping the oxygen from oxygen-rich compounds and combining it with oxygen-poor reduced compounds to generate energy. Some of the earliest bacteria must have fed in this way. It is thus an ancient and very successful feeding strategy that can be put to use to clean the environment. For example, when the element uranium becomes oxidized (as, for instance, by the weathering of uranium-containing rocks), the uranium oxide thus produced is soluble in water and can move through groundwater as a potential contaminant of drinking water. The bacterium *Geobacter metallireducens* is able to live in the ground (hence its genus name *Geobacter*), and has the ability to stop the flow of uranium oxide by stripping the oxygen from the uranium compound and combining this oxygen with an organic molecule such as, say, acetate. In a process known as biogeochemical remediation, scientists have successfully injected acetate as a biostimulant into soil and induced *Geobacter* to metabolize the acetate, strip the oxygen from a threatening plume of dissolved uranium oxide, and render the uranium safely immobile in the ground.

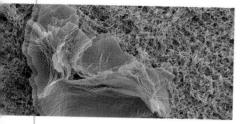

Life and the Crust

There may be good reason to believe that life originated within the crust of the Earth itself. Lynda Williams and her colleagues at Arizona State University have suggested that life may have originated in the vicinity of seafloor hydrothermal vent systems in association with the clay smectite. Smectite and similar clays can shelter and protect organic compounds at temperatures in excess of 570°F (300°C), temperatures that would otherwise destroy these delicate molecules. The Williams scenario is very appealing. Clay minerals such as illite, saponite,

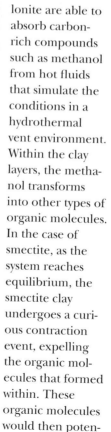

and montmorillonite are able to absorb carbon-rich compounds such as methanol from hot fluids that simulate the conditions in a hydrothermal vent environment. Within the clay layers, the methanol transforms into other types of organic molecules. In the case of smectite, as the system reaches equilibrium, the smectite clay undergoes a curious contraction event, expelling the organic molecules that formed within. These organic molecules would then potentially be expelled into cooler waters outside of the hydrothermal vent system, and would have a good chance of surviving. Thus, the abundant hydrothermal systems likely to have existed on the early Earth's surface might very well have been, as Williams puts it, a "primordial womb" for the origin of life.

EARLY EXPANSION OF LIFE

After the origin event, life swiftly moved into the four primary environments where we find bacteria today: seawater, freshwater, Earth's crust, and land (terrestrial) habitats such as soil. The oldest paleosols (fossil soils) are more than two and a half billion years of age, but this expression of oxygen accumulated in the atmosphere, and bacteria may well have been in soils before this time. Evidence for life in the oceans goes back almost three and a half billion years ago to the Strelley Pool chert in Western Australia. There, C. A. Allwood and colleagues have recognized seven different types of stromatolites (mound-shaped, layered rocks thought to be constructed by bacterial communities that formed microbial mats) preserved intact in an ancient marine sedimentary environment. Evidence for life in the earliest freshwater environments is hard to come

Above: The distinctive plume of a black smoker at a mid-ocean ridge hydrothermal vent. Sulfide minerals in the cloud accumulate as sulfide deposits adjacent to the vents. Oceanic life may go back almost three and a half billion years. Top left: Stereo image of large soil smectite quasicrystal. Clays such as smectite can shelter organic compounds even in extreme heat.

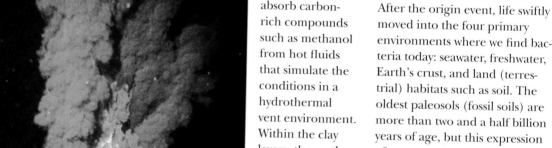

Cross-section of small conical stromatolite at the Strelley Pool chert in Western Australia. The layers beside the conical stromatolite pinch out against its sides, indicating that it grew up above the sediment-water interface.

BACTERIA COOL CLIMATE

As soon as bacteria begin to colonize the surfaces of mineral fragments (in, say, an early soil), the rate of weathering of those mineral fragments begins to escalate dramatically. Organic acids associated with the growth of the bacteria help to dissolve and weather the crystals of minerals such as feldspar. Bacterial metabolism also helps to catalyze mineral weathering reactions such as the Urey reaction, where carbon dioxide is absorbed as certain types of silicate minerals break down into weathering products such as clays. David Schwartzman and his colleagues have argued that such mineral weathering lowered the carbon dioxide levels in early Earth's atmosphere, giving the planet a more habitable climate. More habitable, at least, from the perspective of complex organisms such as ourselves; the earliest bacterial lineages apparently liked it hot and still do.

by at present, but the presence of bacteria in these early habitats can be safely inferred, considering the wide environmental tolerances of what are thought to be the most primitive types of bacteria. Bacteria today live in Earth's crust at depths up to almost a mile and three quarters (2.8 km), and considering the likelihood that life evolved in association with the early Earth's crust, a reasonable hypothesis is that bacteria moved into crustal rocks early in Earth history.

Indeed, evidence has been reported of bacterial fossils billions of years old penetrating the rinds of relatively cool volcanic rocks. Although no one has yet found bacterial fossils in the earliest preserved soils, strong inferences can be made about the presence of bacteria in these early terrestrial habitats based on the concept of biotic enhancement of weathering.

LIFE IN THE CRUST

An interesting question pertaining to life on Earth is the depth to which life extends in the crust. A theoretical limitation to the downward extent of life is associated with the fact that Earth still has much heat in its interior. Some types of thermophilic bacteria can thrive at temperatures up to 235°F (113°C), but temperatures beyond that shock proteins so that they no longer function—and life cannot exist. Nevertheless, bacteria have been retrieved growing beneath Virginia at a 1.75-mile (2.8 km) depth. These bacteria were feeding on ancient organic matter trapped in the surrounding rock, and used metals in the rock such as manganese and iron to oxidize the organic matter and produce energy. Martin Fisk and his colleagues at Oregon State University note that volcanic glass constitutes a significant amount (about one-twentieth) of all oceanic crustal rock. This glass is geochemically unstable, and Fisk has found that this instability is accelerated by sea-crust bacteria that bore into the volcanic glass, mining it for useful metals and other compounds and contributing to rapid weathering of the volcanic glass far below the seafloor.

CHAPTER 2

PLANETARY GEOLOGY

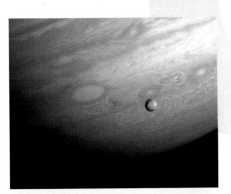

Left: Montage of planetary images taken by spacecraft. From top to bottom: Mercury, Venus, Earth (and Moon), Mars, Jupiter, Saturn, Uranus, and Neptune. The study of the geology of other planets provides scientists with insights into the geology of our own world. Top: Voyager missions to Jupiter resulted in the discovery of active volcanism on the moon Io—the first observation of active volcanoes on another body in the solar system. Bottom: The stunning false-color view of Saturn's Hyperion reveals crisp details across this strange, tumbling moon's surface.

Planetary geology is the study of planets and their satellites, or moons. From orbiting telescopes to Martian rovers, vast amounts of new geological data are being collected on the planets in our solar system. This data may give insights into geology back home. Most planetary research relies on new geoprocessing tools, and great advances are being made in the geological study of planetary surfaces thanks to a wide range of space probes and other orbiting devices. As space probes explore farther into the solar system, a particularly exciting aspect of the quest is the detailed images and other data coming from the moons, asteroids, and comets. For example, until recently it was thought that Jupiter's Io was the only moon to show recent volcanic activity. *Voyager 2* imagery of Enceladus, however, confirmed that on this moon of Saturn a continuously gushing icy plume shoots up 620 miles (1,000 km) above the surface. Images of Neptune's Triton reveal geysers there, as well. Hyperion, an oblong satellite also of Saturn, has a spongelike texture consisting of radiating pores that make this moon resemble a giant, floating block of pumice. These strange images are raising many more questions than they answer, symptomatic of the excitement that planetary geology continues to generate.

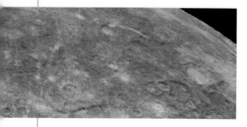

Mercury

The planet closest to the Sun, Mercury has a surface characterized by refractory materials that can take the heat. Mercury is distinguished both by its relatively small size (diameter 3,030 miles, or 4,880 km); no known satellites; and its eccentric orbit that goes from 28.6 million miles (46 million km) at its nearest

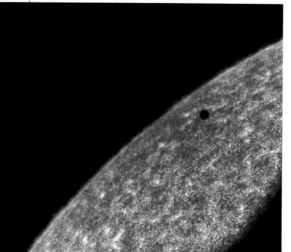

approach to the Sun to 43.5 million miles (70 million km) at its most distant. A planet of extremes, Mercury has the greatest temperature fluctuations of any planet in the solar system, ranging from -297° to 800°F (-183° to 427°C). Persistent variations in Mercury's orbital parameters puzzled astronomers until they were explained, in

a spectacular example of a successful scientific test, by an application of Einstein's theory of general relativity. A final surprise from this planet is evidence for water ice at the poles; radar data taken from the deep shade of north polar craters disproves the generalization that the surface of Mercury consists only of refractory materials.

AKIN TO EARTH

Mercury is the terrestrial, or Earth-like (in terms of crust and core composition), planet closest to the Sun. Because of density similarities to Earth, Mercury's interior composition is probably very similar. Like Earth, Mercury is thought to have a dense metallic (and probably partially molten) core 1,058 miles (1,700 km) in diameter, surrounded by a silicate mantle and crust a mere 340 miles thick (550 km), which gives Mercury a proportionally much larger core than Earth. The core evidently generates a weak magnetic field. Early attempts to understand the surface geology of Mercury met with challenges because it is hard to get a clear view of the planet's

surface with a medium-sized telescope. Italian astronomer Giovanni Schiaparelli (1835–1910) swore off tobacco, coffee, and whiskey in 1881 to ensure the clearest vision possible in his pioneering attempts to map Mercury's surface. Parisian astronomer Eugène Antoniadi (1870–1944) provided more detail in a 1934 map of Mercury, and subsequently a pair of major ridges imaged by the *Mariner 10* space probe are called Antoniadi and Schiaparelli to honor their

Left: The minuscule black dot in this composite picture of the Sun is the planet Mercury as it passes in front of the Sun. Bottom: Mosaic of Mercury's crater-filled surface was taken during Mariner 10's *approach. Top left: This color composite of Mercury was formed to especially highlight differences in opaque minerals (such as ilmenite), iron content, and soil maturity.*

NASA rendering of Mariner 10. *The ten Mariner missions of the 1960s and early 1970s were designed to visit our nearest neighbors in the solar system. Seven were successful, reaching planets Venus, Mars, and Mercury.*

with another body from outer space). The greatest of these is the 800-mile-diameter (about 1,300 km), mile-deep Caloris Basin. Although the surface of Mercury is very ancient and never developed plate tectonics, huge escarpments associated with the thrust faults provide evidence that the planet has indeed shrunk. Featureless plains on Mercury are probably the result of either impact ejecta fallout or ancient lava flows.

early work on the geology of Mercury. The mapping effort continues, and the MESSENGER (MErcury Surface, Space ENvironment, GEochemistry, and Ranging) space probe is slated to begin detailed mapping of the planet surface in early 2008. Temperatures are so great this close to the Sun that

MESSENGER must be shaded by a heat-deflecting ceramic shroud; temperatures drop off fast enough in the shade to allow water ice to exist. The information expected from MESSENGER will be of great interest, as there is evidence from the earlier imagery (in the form of suspected thrust faults) that the planet has undergone considerable shrinkage, with loss of one-half kilometer of the planet's radius.

MERCURY REVEALED

Mariner 10's images of Mercury show a planet surface pockmarked by impact events, with streams of ejecta in a variety of different albedos (whiteness or reflectivity) radiating from the astroblemes (the scars left on a planetary surface by the impact

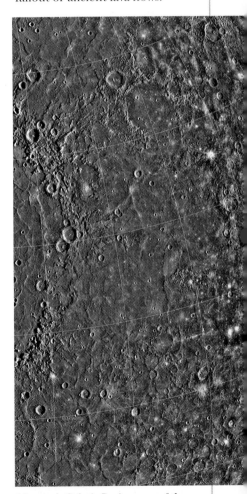

Mercury's Caloris Basin—one of the largest in the solar system—exceeds 800 miles (1,300 km) in diameter.

Venus

Venus is farther from the Sun than Mercury and comes closer to the Earth than any other planet, but its surface temperatures are nevertheless extremely hot. Venus is hotter than Mercury due to the runaway greenhouse atmosphere, the result being a surface temperature far more stable than Mercury's. No life seems to be present to maintain this equilibrium. The great Russian geologist, poet, and astronomer Mikhail Lomonosov (1711–65) first detected the Venusian atmosphere in 1761. This cloudy atmosphere accounts both for Venus's notable brightness (it can be seen clearly in broad daylight if you know where to look, and can cast shadows at night) and for the impossibility of seeing anything of the solid surface of the planet from an earthbound light telescope. Like Mercury, Venus lacks any satellites. The Venusian year lasts 227.4 days in Earth reckoning. With a radius nearly 95 percent that of the Earth, Venus has been called Earth's "sister" planet.

RUSSIAN RESEARCH

Lomonosov's pioneering research later inspired both Russian interest in planetary atmospheres (as developed by geochemist Vladimir Vernadsky) and the Russian series of space probes (*Vega* and

Below left: Venus cloud tops viewed by Hubble. Venus is covered with clouds made of sulfuric acid, rather than the water-vapor clouds found on Earth. These clouds permanently shroud Venus's volcanic surface, which has been radar mapped by spacecraft and from Earth-based telescopes. Below right: Hemispheric view of Venus. The Magellan *spacecraft imaged more than 98 percent of the planet. Top left: The desert surface of Venus taken by the* Venera 13 *lander.*

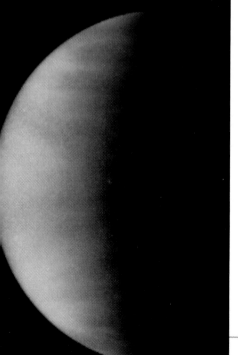

Venera) that reached the surface of the planet. If Venus ever had oceans, they were lost to dissipation long ago. Many scientists are now pointing to Venus as a cautionary tale for the residents of Earth, an example of runaway planetary heating partly a result of the accumulation of compounds of chlorine in the upper atmosphere. Venus is of course closer to the Sun, but could such a fate ever envelope the surface of Earth? Carbon dioxide is accumulating in Earth's atmosphere, and the atmosphere of Venus is 97 percent carbon dioxide. The 800°F surface temperature and the crushing atmospheric pressure limited the functional life of the Russian landing craft to minutes. The great atmospheric pressure on the surface of Venus approaches 100 times the atmospheric pressure at the Earth's surface. A continuous

acid rain known as virga adds to the inhospitability of the Venusian surface.

CLUES FROM SPACE PROBES

Virtually nothing was known of the surface morphology of Venus until radar analysis from Earth in the 1960s revealed a prominent mountain chain five miles (8 km) high. These mountains were called Maxwell Montes for James Clerk Maxwell, and the lowlands to either side of the mountain range were referred to as Alpha Regio and Beta Regio. This was the extent of our knowledge of the Venusian surface until the arrival of the *Magellan* space probe in 1990. In the four years that *Magellan* orbited the planet, an almost overwhelming amount of landform imagery was sent back showing the surface of what is essentially a volcanic planet.

Most of the surface of Venus is less than a half billion years old, as lava flows have covered over most of the older rock. The volcanic features in view, however, are remarkable in their diversity, with domes surrounded by concentric ring structures called coronae, flat-topped volcanoes referred to as pancake domes or farras, numerous small volcanoes that contrast strongly with the smooth lava plains, elevated plateaus faulted into geometric tile patterns (tesserae), and strange spiderlike features.

Venus's cloud cover obscures any view of the surface in this image from the Pioneer *Venus orbiter.*

TECTONICS ON VENUS?

In the opinion of some, Venus has experienced episodes of crustal subduction, suggesting that the planet undergoes a process of plate tectonics similar to that on Earth. Not everyone agrees with this assessment, and sometimes lack of tectonics is cited as a difference between Earth and Venus. Probably some sort of intermittent tectonic processes have operated on the crust of Venus.

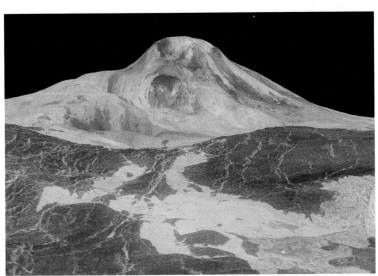

The volcano Maat Mons is displayed in this three-dimensional perspective view of the surface of Venus. Lava flows extend for hundreds of kilometers across the fractured plains shown in the foreground, to the base of Maat Mons.

Mars

This enigmatic planet has ice caps, evidence of ancient oceans, and glaciers possibly associated with subsurface gas hydrate deposits. In terms of surface geology, Mars, along with Venus, is the most Earth-like of the other planets of the solar system. This was not always thought to be the case; because of its splotchy surface albedo, Mars was once thought to be covered in canals and, in the seventeenth century, Swiss mathematician Matthias Hirzgarter illustrated Mars as a massive triangular rock. The points of his triangle might today correlate with the regions of Mars known as Syrtis Major Planus, Mare Serpentis, and Tyrrhena Terra. Percival Lowell, in his 1906 book, *Mars and its Canals*, was convinced that canals built by the water-starved remnants of a dying alien civilization covered the surface of Mars. The canals proved to be illusory, with the sole exception of the enormous fault-bounded fissure in the Martian surface called Valles Marineris. This impressive linear feature roughly parallels and is just south of the equator. Its various sections, from west to east, are named Tus Chasma, Malas Chasma, Coprates Chasma, Capri Chasma, and Eos Chasma. More depressions, some more or less parallel to Valles Marineris, occur just to the north: Hebes Chasma, Ophir Chasma, Candor Chasma, Juventae Chasma, and Ganges Chasma. Where canals were lacking, chasmas stepped into the breach.

MARTIAN TOPOLOGY

Mars has an interesting surface geology that is divided into two main regions: a region of older highlands south of the equator and less ancient plains to the north. The southern highlands are heavily cratered, with craters named in honor of Chamberlin,

Above: A close-up of the "red planet," Mars, taken by NASA's Hubble Space Telescope. Top left: A magnified look at Martian soil shows coarse grains sprinkled over a fine layer of sand. The spherical rocks may have been formed by a variety of geologic processes, including the cooling of molten lava.

Lowell, Darwin, Wallace, and Newton. These craters attest to the great age of the southern highlands; they date back to the 4.5- to 3.8-billion-year interval of late but heavy planetesimal bombardment of the early planet. The plains (singular: *planitia*) of the north are quite topographically low with respect to the highlands, so low, in fact, that the north pole of Mars is four miles closer to the equator than is the more ancient south pole terrane. As if to compensate, rising up from the Amazonis Planitia in the Northern Hemisphere is Olympus Mons, the greatest volcano in the solar system.

Its great height may be attributed to two geological factors. First, because of the thin atmosphere and extremely limited hydrologic cycle, the erosive forces on Mars are decidedly feeble. Martian air can raise the occasional dust devil, but these have not been much of an erosive threat to Olympus Mons. Second, the lack of plate tectonics on Mars left the volcanic vent of Olympus Mons stationary during the entire eruptive interval, leaving the magma generated thereby to accumulate all in one place. Conversely, the southern highlands sport their own plains such as the Argyre Planitia and the enormous, deep Hellas Planitia. Both of these resemble huge astroblemes, or impact scars. Hellas Planitia occurs right below Hirzgarter's triangle, which consists of Syrtis Major Planum and Tyrrhena Terra, and is thus visible from a great distance.

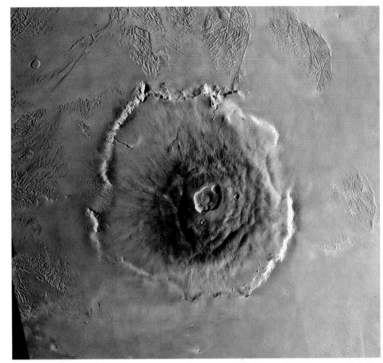

Olympus Mons is a shield volcano 374 miles (600 km) in diameter and 16 miles (26 km) high. Hawaii's Mauna Loa, Earth's largest volcano, also a shield volcano, measures only 6.3 miles (10 km) high and 75 miles (120 km) across.

A swath of Valles Marineris. The "Grand Canyon" of Mars is about 2,500 miles (4,000 km) long and up to 4 miles (6.5 km) deep. The Earth's Grand Canyon is less than 500 miles (804 km) long and 1 mile (1.6 km) deep.

Mars (continued)

MARTIAN ATMOSPHERE

Mars's carbon dioxide–rich atmosphere is decidedly wispy in comparison to that of Earth. Atmospheric pressure on Earth outstrips that of Mars by a factor of 100. More than 30 percent of the Martian atmosphere is seasonally lost to the carbon dioxide ice that forms at both poles. The ice caps occur in the Planum Boreum and the Planum Australe, respectively. One of these polar caps may have been spotted as early as August 13, 1672, by Christiaan Huygens, whose sketch of the planet showed a circular bright region at its lower end. In addition to dust devils, Martian air carries fine dust particles in suspension, giving the Martian sky its characteristic salmon pink tint. *Mariner 9* images of Tyrrhena Terra show elongate dust accumulations deposited in the lee slope of crater rims, indicating that winds can blow on Mars with prevailing directions, and carry and deposit sediments while doing so.

EVIDENCE OF WATER?

The early interval in Martian history is called the Noachian, with the implication that there was a once significant amount of water on the Martian surface. The *Pathfinder* landing rover

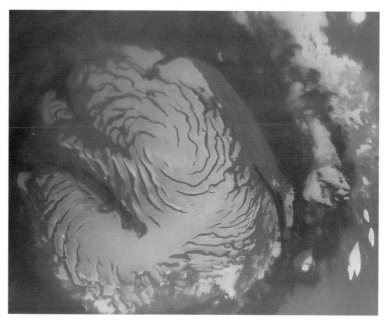

Top: The Martian North Pole in summer, covered by a cap of water ice. In winter it is covered by a layer of solid carbon dioxide. Bottom: "Windlanes" on the Martian surface are caused by the wind transporting loose sand fragments that impact on the bedrock, slowly removing parts of the surface, like a sandblaster. Top left: The Martian rock "Barnacle Bill," which scientists found to be andesitic, indicating that it is a volcanic rock.

Intended to gather information on planetary physics, geology, planetology, and cosmology, Mariner 9 *was the first spacecraft to send back data from Martian orbit, including surface images of Mars's two moons, Deimos and Phobos.*

has brought back evidence of water-deposited sediments bearing strange sulfide minerals tentatively identified as jarosite. Sinuous channels imaged by orbiting space probes on the Martian surface are suggestive of river meanders and imply forces of flowing water. Some crater depressions are thought to have hosted lakes, and even deltas have been identified forming in these crater depressions. Extreme caution must be exercised in such identifications, however, because it can be difficult to distinguish between, say, a water-deposited delta and a lava delta. The imagery from MOC and other sources has really underscored the need for additional, direct reconnaissance of the Martian surface.

Convincing evidence for water flow is available. The 1972 *Mariner 9* images revealed the Nirgal Vallis to be a river channel with meander bends on its downstream part and branching tributary streams up-gradient. Some evidence for water on the Martian surface takes the form of what is interpreted as collapsed ground, disrupted regions that were apparently destabilized by release of water as the result of the decomposition of permafrost or gas hydrates.

WAS THERE LIFE ON MARS?

Methane has been detected in the Martian atmosphere, and scientists are puzzled as to the source of this hydrocarbon. Where is the methane in the Martian atmosphere coming from? The Allan Hills meteorite, blasted off the Martian surface about 15 to 18 million years ago in an impact event and discovered in Antarctica in 1984, is one of the oldest rocks known in the solar system at 4.5 billion years old. It has other curious geological properties, including magnetite crystals that some scientists say could only have been formed by bacteria (as opposed to inorganic processes). Although the evidence was far from conclusive, this single rock fueled debate over whether Mars had life. The detection of methane in the Martian atmosphere added fuel to the discussion. Some scientists speculated that it was derived from subterranean Martian microbes. Decomposition of underground gas hydrates may again be a factor. The case, however, is far from solved.

Moons of the Giant Planets

Jupiter and Saturn each have a host of orbiting moons. These moons are numerous; Saturn alone possesses at least 35; Jupiter has somewhere around 60. The surface of the Saturnian moon Titan harbors an icy bedrock and may also host tholins, chemicals soluble in water that serve as a starting point for a variety of organic compounds. The Jovian moon Europa is thought to have an active surface, constantly renewed, because it is the smoothest object in the solar system.

JOVIAN MOONS

Of its large moons, the closest to Jupiter is the red moon Io, which orbits the vast planet in less than 48 hours. Io has abundant evidence for volcanism on its surface. Next comes Europa, thought to have a sea beneath its smooth

Artist's depiction of the area surrounding the Huygens *landing site. The European Space Agency's probe reached the upper layer of Titan's atmosphere and landed on the surface after a 2-hour and 28-minute parachute descent.*

surface. Both Io and Europa are close enough to Jupiter to feel powerful tidal forces; any closer in and they would cross the dangerous threshold of the Roche limit (beyond which moons are unstable) and be torn apart by the tidal forces. Past Europa is Ganymede, the largest moon in the solar system, bigger in fact than Pluto or the planet Mercury. Ganymede feels

Left: Jupiter, Io, and the Galileo *space probe. Jupiter's moon Io is the most volcanically active body in the solar system. Top left: Cassini's view of the pale, icy moon Dione, with Saturn visible in the distance.*

enough of Jupiter's tidal friction to maintain a partly molten core capable of generating a planet-like magnetic field. Next out is Callisto, far enough from its planet to be free from tidal heating. Callisto is heavily cratered due to impact battering.

SATURN'S SATELLITES

Saturn's largest moons are Mimas, Enceladus, Tethys, Dione, Rhea, Titan, and Iapetus. In 2005 *Cassini-Huygens* imagery showed that Enceladus, with a 310-mile (500 km) diameter, releases a plume of icy particulates that shoots into space

Rendering of the bright, icy surface of Saturn's moon Enceladus, which illustrates an ice geyser projecting a jet of vapor into space.

a distance equal to its diameter. Titan is a huge orange ball, much larger than Enceladus at 3,200 miles (5,150 km) diameter, and rivaled in size only by Jupiter's Ganymede at 3,270 miles (5,260 km).

In what was arguably the most interesting spacecraft landing since the Apollo moon missions, on January 14, 2005, the European Space Agency's probe *Huygens* descended to the surface of Titan. On its way down, it sent images of what seemed to be an eerily familiar landscape. A panoramic mosaic image showed a ridge cut by more than a dozen channels or fissures. To the north and west of this landscape are seen dark, bifurcating channels that are not unlike drainage channels on Earth. Greatly assisting the imagery effort, the presumed channels have dark beds that contrast vividly with the light-colored terrane through which they run. These Titan channels look even more familiar and Earth-like than many of the channels on Mars, and it is postulated that a methane rain formed

the channels. This is a very interesting concept: methane rain eroding into, and forming rounded rocks from, an icy terrane. Taking the comparison to Earth one step further, a number of the dendritic channels drain down into what appears to be the shoreline of a major river channel. Clearly on Titan there is some kind of strange variation on the hydrologic cycle of Earth, with both intriguing similarities and illuminating differences.

Once on the surface, the *Huygens* lander sent back images of a stone-strewn surface showing rocks up to 5.91 inches (15 cm) in diameter, presumably composed of water ice. The stones rest on a bed of darker, fine-grained material and show scour channels where the fine material has been eroded by some sort of flow process. The stones are rounded, much like rocks on Earth, indicating that they have been subjected to transport and rounding. Indeed, the Titan rocks look very similar to those on Earth that one might find in a dry lake or riverbed.

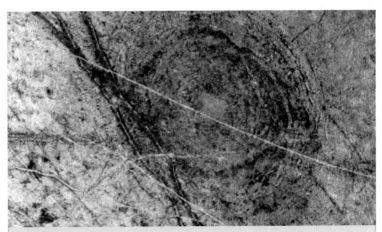

The bull's-eye pattern appears to be an impact scar that formed at the surface minutes after an asteroid or comet slammed into Europa.

IMAGING EUROPA

The surface of Europa is extremely smooth (one of the smoothest surfaces in the solar system, in fact) and is crisscrossed by low-albedo linear features that evidently represent healed fractures in an icy crust. These fractures, and areas that look in images like frozen pack ice, indicate that Europa has a thick icy crust supported by a great water reservoir.

Saturn and Jupiter

Galileo Galilei's sketch of Saturn in 1609 and especially his time-series sketches of the Jovian moons in 1610 transformed planetary observations into a science. Planetary research developed rapidly from this start (and diverged from astrology; Galileo was a practicing astrologer), and in

1675 Giovanni Domenico Cassini had identified four of Saturn's satellites and also the curious gap in Saturn's rings now known as the Cassini Division.

THE GAS GIANTS

Strictly speaking, Jupiter and Saturn lack surface geology because they do not have a

solid surface or rocks as we understand them. Because most of the mass is in fact liquid hydrogen, it is not truly proper to call these planets gas giants. Jupiter has no solid surface, for in its dense atmosphere (curiously enriched in noble gases such as xenon, krypton, and argon) solids are absent except, possibly, for a rain of tiny diamonds formed by the crushing of methane under the tremendous pressure. Jupiter has been called a failed sun, hot on the inside (as much as 43,000°F, or 24,000°C) but not hot enough to sustain the fusion reaction that powers a star. The pressure inside Jupiter is so great that the atmosphere is compressed into a layer of liquid hydrogen, and the outer atmosphere (tinted a variety of colors by sulfur and other compounds) is in continuous flux, playing host to thunderbolts hundreds of times more powerful than anything on Earth. The appropriately named *Galileo*

Left: A composite of the Jovian system includes, from top to bottom, Io, Europa, Ganymede, and Callisto, and the edge of Jupiter with its Great Red Spot. The Great Red Spot is a storm in Jupiter's atmosphere that is at least 300 years old. Top left: Infrared image of Saturn's rings.

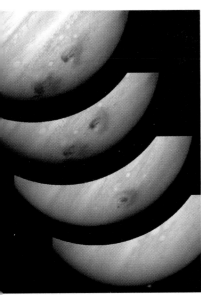

Hubble images chronicle the results of Jupiter's collision with chunks of comet Shoemaker-Levy 9. In the first from the bottom, five minutes after the collision, no surface changes are visible, yet less than two hours later, a plume of dark debris can be seen. Two impact sites are visible a few days later. In the final image, three impact sites are visible, the newest near the bull's-eye-shaped region.

space probe identified the presence of water vapor in the Jovian atmosphere. Jupiter's liquid hydrogen sea has a metallic character and is responsible for generating a magnetic field so powerful that it remains strong out to Saturn's orbit.

COSMIC COLLISIONS

When comet Shoemaker-Levy 9 smashed into Jupiter in July 1994, the force of the impacts

(the comet had fragmented into 21 pieces) created dark smudges that became visible in Jupiter's atmosphere, easily seen by Earth-based telescopes. By the end of August 1994, the continuous weather of Jupiter had all but effaced any evidence of the impact.

Saturn's ring was considered a single solid ring until recognition of the Cassini Division divided the structure into an outer ring, B, and an inner ring, A. This process of dividing up Saturn's ring reached its climax in 1857 when James Clerk Maxwell's calculations demonstrated that the ring must be composed of innumerable solid particles. What was the origin of this curious set of rings?

The particulates of the rings are gravitationally attracted to

one another, and would long ago have formed a moon of Saturn, but for the fact that they are too close to the giant planet. The rings fall within the Roche limit, where the tidal forces generated by Saturn's gravitational field pull apart any hopeful aggregations of ring rocks. The leading inference for the genesis of Saturn's rings is that they are formed of particles that represent the remains of a captured asteroid or other body that was torn apart due to its position within the no-man's-land Roche zone.

The Cassini spacecraft transmitted radio signals through the rings of Saturn, making it possible to produce this image showing the density of ring particles.

Outer Planets and Company

Comets, asteroids, and planets with their moons occur in the outer reaches of the solar system. The surfaces of all these objects provide clues to the processes of planetary formation. The existence of Neptune was predicted in 1846 by Urbain Jean Joseph Le Verrier as a means of explaining irregularities in the orbit of Uranus, and ever since, the two planets have been linked in terms of their scientific study.

LINKED PLANETS

Like Jupiter and Saturn, Uranus and Neptune have no solid surface. For Uranus, the demarcation between its atmosphere and the planet proper is arbitrarily defined as the spherical isobaric surface where the gas pressure reaches one bar (one Earth atmosphere). Uranus's amazing blue hue is due to

methane in its outer atmosphere. This methane, which constitutes 2.3 percent of Uranus's atmosphere, absorbs red light in a greenhouse gas fashion (red as well as infrared), leaving the planet with its distinctive turquoise.

Like Saturn, both Uranus and Neptune have rings, the probable remains of satellites that passed too close to the planet and fell within the Roche limit where destructive tidal forces tore them apart. Uranus's ring system is elliptical in shape, whereas Neptune's rings are very nearly circular. Neptune's rings have been named Adams, Le Verrier, and Galle, with the Adams ring being further subdivided into three arcs curiously named Liberty, Equality, and Fraternity.

NEPTUNE'S MOONS

Neptune's largest moon, Triton, is unique in the solar system because it rotates in the wrong

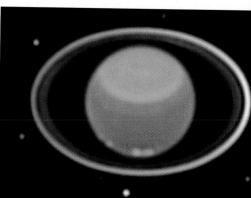

Infrared image of Uranus. Its rings are extremely faint in visible light but quite prominent in the near infrared.

direction with respect to all the other larger moons. This retrograde orbit has led to informed speculation that Triton is a captured satellite, possibly originally derived from the distant Kuiper Belt. Triton is very similar in size and composition to Pluto, and this has led to the alternate hypothesis that Triton originated as a dwarf planet much like Pluto before its capture by Neptune. Theoretical considerations even suggest that Triton may have once had a satellite, comparable to Pluto's moon Charon, immediately before its gravitational capture. Compared to the other gas planets, Neptune lacks a large set of moons, leading to the hypothesis that the appearance of Triton led to orbital destabilization and eventual loss

Left: A false color photograph of Neptune, taken from Voyager 2 *images. Top left: Uranus's five large moons, from left, in order of increasing distance from the planet, they are Miranda, Ariel, Umbriel, Titania, and Oberon.*

Neptune and one of its moons, Triton. Triton is the smaller crescent and is closer to viewer.

of Neptune's primordial moons. Perhaps some of these hypothetical bodies were released toward the inner planets of the solar system where they acted as planetesimals with potential to cause scarring impacts. Triton's retrograde motion is causing it to spiral inward toward Neptune, and calculations indicate that in about two billion years, Triton will pass within the Roche limit and be destroyed by tidal interactions with Neptune. Assuming that the fragments are not swallowed up by Neptune's atmosphere, they might very well organize themselves into a ring, adding to Neptune's existing rings. Almost as if in anticipation of its dismal fate, Triton is the coldest object ever detected in the solar system.

PLUTO AND KUIPER BELT OBJECTS

Pluto, only 1,413 miles (2,274 km) in diameter, has as yet not been visited by one of our space probes. Pluto's status in the Solar System was downgraded in August of 2006 from planet to Kuiper Belt object, or dwarf planet. Although Pluto has

a spherical shape, it has not gravitationally captured the other objects in its orbital neighborhood and thus is not qualified for planet status. Other dwarf planets in our solar system include the largest asteroid, Ceres, and an object, larger than Pluto, officially known as UB 313, but informally as Xena. The discovery of Pluto's moon Charon in 1978 allowed calculation of Pluto's mass. In addition

to Pluto's small size (seven moons are larger: Ganymede, Titan, Callisto, Io, Earth's Moon, Europa, and Triton), a further blow to Pluto's status as planet is its elliptical orbit that passes within the orbit of Neptune for a significant part of its path. At its farthest extent, Pluto's orbit takes it 4.6 billion miles (7.4 billion km) from the Sun, nearly 50 times the distance from the Sun to Earth.

Pluto and its moon Charon are locked into a dyadic orbit that keeps the same part of each body facing the other. Binaries such as this are common in the Kuiper Belt, and it is reasonable to consider the Pluto-Charon binary as belonging to the class of Kuiper Belt objects.

The clearest view yet of the distant Pluto and its moon, Charon.

AN ELUSIVE OBJECT

Because no spacecraft has yet visited Pluto, assessing its surface has proved to be a challenge similar to the earlier challenge of mapping the surface of Mars when all that astronomers had to examine it with were earthbound telescopes. Splotches of varying albedo have been observed on Pluto's surface (the best image was taken during an eclipse of Charon), but we must await the visit of the space probe *New Horizons* before we can expect more detail. *New Horizons*, launched in January 2006, will briefly visit Pluto and Charon in July 2015.

CHAPTER 3

MINERALS, ROCKS, AND THE CRUST

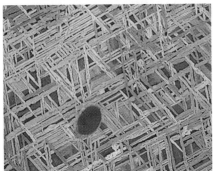

Left: A clearly visible contact between the lower basaltic and upper felsic ash units in the wall of the West Rota caldera, Mariana Arc System. The outcrop in view is approximately 6.6 feet (2 m) in height. Top: A small phreatic eruption from Mount St. Helens in the spring of 1980, shortly before the deadly May 18 blast. Bottom: Widmanstätten structure in the iron-nickel minerals taenite and kamacite, with taenite containing much more nickel than kamacite. These minerals are thought to approximate the composition of Earth's core.

The nuclear fusion that took place inside of stars, and the dramatic explosions that end their lives as supernovae, both form and distribute heavier elements such as iron, carbon, magnesium, manganese, silicon, uranium, and aluminum, the elements that came together to form minerals that created, first, planetesimals, and subsequently, the Earth's crust. Rocks are generally composed of interlocked agglomerations of a variety of minerals. The types of minerals present determine the rock type, and each rock type (igneous, sedimentary, or metamorphic) can be classified according to the type of mineral crystals (and their sizes) it contains, with the exception of types such as volcanic glass, where its elements are not organized into ordered crystal structures.

Earth's crust is composed primarily of two types of igneous rocks. The darker and denser of these two is referred to as mafic, for its elevated content of magnesium (chemical symbol Mg) and iron (Fe). Mafic rocks include dark volcanic and igneous rocks such as basalt and gabbro, respectively. Felsic rocks are also igneous, but are lighter and less dense because of their higher proportion of the lighter elements silicon and aluminum. Mafic rocks compose seafloor crust; felsic rocks are the main constituents of continental crust.

Olivine to Quartz

The silicate minerals of Earth's crust can be organized and classified according to the amount of connectedness between the silica tetrahedra that compose these minerals. The basic unit of silica, the silica tetrahedron, is a pyramid-shaped molecule of silicon and oxygen that can form the backbone of critical crustal minerals such as pyroxene and amphibole. These tetrahedra can link together to form chains (pyroxenes such as jadeite), double chains (amphiboles), sheets (mica), tectosilicates (feldspars such as sanidine), and even a diamondlike framework

(quartz). This system of crystal classification has a beauty, regularity, and order that is all its own, and can even be related to the sequence of crystallization of minerals from a lava or magma. This sequence of crystallization is known as Bowen's Reaction Series, and the general message of this series is that the

Left: Radiating quartz with well-formed crystal faces. These formed in a silica-rich fluid-filled cavity in a large rock body. Above: Olivine particles from the Green Sand Beach in Hawaii. Top left: Wollastonite from Isola d'Elba, Italy. The chemical breakdown of wollastonite absorbs carbon dioxide from the atmosphere.

less ordered silicates, that is, those with less ordered or even isolated silica tetrahedra such as olivine, are the first to form in a cooling magma.

As the magma continues to cool and the crystallization moves to completion, successively more ordered types of silicate minerals form, from chain silicates to double-chain silicates to sheet silicates (micas) to, finally, tectosilicates such as quartz or essential pure silica where the tetrahedra are organized into a very strong three-dimensional array that resembles the powerful framework bonding of carbon in diamond. Since quartz is the last mineral to form in, say, a granite, its crystals tend to be misshapen because, as it grows, it has to fit around the less-ordered silicate minerals such as feldspar that formed early on in Bowen's Reaction Series. Interestingly, and again in general, the less-ordered silicate minerals such as olivine are more susceptible to weathering, and the more ordered silicate minerals such as quartz are more resistant to weathering and tend to be more likely to end up as a weathered-out clast (such as a sand grain of quartz) in a sedimentary rock such as a sandstone.

SILICATE VARIATIONS
Within each class of silicate minerals there are interesting variations. Consider, for example, the single-chain silicate minerals such as pyroxene, wollastonite ($CaSiO_3$),

rhodonite ($MnSiO_3$), and pyroxmangite ($MnSiO_3$). The single silica chain in pyroxene consists of a sequence of alternating silica tetrahedra that are bonded end-to-end and alternate left to right in a chain pattern. The most basic unit of the chain is a sequence of two tetrahedra (5.2 angstroms in length) that is repeated along the chain in both directions. In wollastonite, the silica chain is repeated in a pattern that has two tetrahedra side to side, then one projecting outward, two tetrahedra side to side, and another projecting outward, and so on, to form the chain. The basic pattern in wollastonite may be reduced to a minimal unit of three tetrahedra (two end to end, one to the side) that again repeats in both

The mineral rhodonite ($MnSiO_3$), with its characteristic rose color, is an example of a single-chain silicate.

directions. The wollastonite unit is 7.2 angstroms long.

In rhodonite, the basic unit of the chain is a twist of five tetrahedra 12.2 angstroms in length that extends in both directions to form the chain. And in pyroxmangite, the basic unit is even more complex, consisting of seven tetrahedra forming a repeating unit that is 17.4 angstroms in length.

Characteristic tartan pattern seen in the mineral microcline as viewed in a petrographic thin section. The pattern permits recognition of this mineral even when it occurs as a weathered clast in sedimentary rock.

PATTERNS IN ROCKS
Microcline is a feldspar that belongs to the group of tectosilicate minerals. Rather nondescript as a hand sample, it transforms into an amazing tartan pattern when viewed in a petrographic thin section. This pattern is very helpful to both igneous and sedimentary petrologists because this type of feldspar can be identified in thin section at a glance. It is one of the feldspar varieties more resistant to weathering, and thus does tend to end up as a component of sedimentary rocks such as sandstone, even when the sandstone is composed primarily of quartz grains. In some mature sandstones, rare grains of microcline are the only feldspars present.

Mafic Rocks

Mafic rocks are entirely igneous; the main differences between the various types relate to the sizes of the crystals that compose these rocks. For example, basalt and gabbro have essentially the same compositions but differ based on the sizes of their constituent crystals. Gabbro forms from a magma that cools slowly beneath the surface of the crust; hence, its crystals have time to grow to relatively large size. Basalt, on the other hand, forms closer to the surface (or, in the case of a mafic lava flow, right at the surface) where cooling rates are faster.

Basaltic magma therefore crystallizes more rapidly and the crystals of basalt are, as a result, smaller than those of gabbro.

VOLCANIC GLASS

The cooling of magma destined to form basalt has some interesting aspects that are worthy of discussion. If such a magma is chilled very quickly, no crystals have a chance to form and instead a volcanic glass is formed. This glass, often encountered on the surface of lava flows, can have the same

chemical constitution as basalt, but its silica tetrahedra are highly disorganized, thrown together in a helter-skelter mixture that has more in common with a viscous liquid than it does to a crystalline solid. Indeed, it has a chaotic

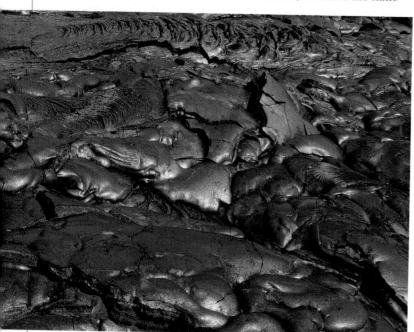

Left: Pahoehoe, a smooth ropy lava, formed where lava is erupted as a slow-flowing mass near Mauna Ulu, Hawaii. Such glassy lavas, like all glasses, are solids having the disorganized molecular structure of liquids rather than the organized, crystalline structure of magmas that cooled more slowly. Above: Deformed gabbro from the Atlantis Massif at the Lost City Vent Field. Many of the rocks recovered from near the top of this submarine mountain are highly deformed from stresses of extended periods of fault activity. Top left: Contraction of cooling lava formed these basalt columns in Turkey's Boyabat Province.

These ancient pillow basalts either erupted from an underwater volcanic vent or flowed into water from a fissure on land. Tectonic activity subsequently uplifted the lava deposit.

because they have the shape of pillows thrown together on a bed as seen in a typical geological cross-section exposure. A typical pillow basalt has a glassy rim or rind surrounding a core of very fine-grained basalt. Typically, the fine-grained part of the rock will show, on a freshly broken surface, tiny white flecks that represent small plagioclase feldspar crystals.

When mafic lava is erupted onto land, it can flow for some distances to form lava sheets that cool and crystallize to form flood basalts. These sheets cool over the course of days or even weeks, and as they cool they contract and tend to fracture into a regular pattern known as columnar jointing. When well-developed, columnar jointing in flood basalt forms beautiful, hexagonal prismatic columns that look like they have been carved and smoothed by human hands.

structure that resembles that of glass used in our windowpanes. Incidentally, the volcanic glass known as obsidian, although usually dark in color, has a lot of silica and is therefore a felsic rock with a composition similar to that of granite or its fine-grained counterpart, rhyolite.

that were chilled and hardened before they had a chance to spread out into lava sheets as they often would do if erupted under air. This is so because water can remove heat faster from the lava than can air in most conditions. These chilled lava tubes are called pillows

Obsidian, a black volcanic glass, showing its characteristic conchoidal fracture. Obsidian was a favorite raw material for Paleolithic tool manufacture.

PILLOW BASALTS

An interesting combination of glassy and crystalline textures can give geologists clues as to whether or not a particular mafic lava flow was erupted into a body of water such as a lake or sea. When such magma is erupted into water, it forms what is called a pillow basalt. Pillow basalts are actually solidified tubes of lava

Felsic Rocks

Like mafic rocks, felsic rocks come in two varieties: coarse-grained and fine-grained, depending on how fast the felsic magma that form them was allowed to cool. A typical felsic magma that cools slowly will form a granite. Granites have individual crystals that are visible to the naked eye, and will include such mineral types as feldspar, biotite (a mica), and quartz. If a magma that would otherwise form a granite is allowed to cool quickly, it will form a fine-textured rock called rhyolite. Rhyolite is a crystalline rock rather than a glass, but most of its crystals will be too small to be visible to the naked eye. Thin sections mounted to glass can be viewed by means of a petrographic microscope. Occasionally, rhyolites will contain scattered larger crystals such as clear, elongate sanidine feldspar crystals.

GRANITIC ROCKS

Granites typically form at depth in the crust, a part of bulbous magma chambers that, when they cool, form gigantic spheroidal masses of rock known as plutons. Plutons start out beneath the Earth's surface, but as the rocks overlaying them are subjected to erosion, they tend to be exposed at the surface over time. Many of our greatest and most beautiful mountain ranges, such as the Sierra Nevada, the Andes, the Rocky Mountains, the Adirondacks, and the White Mountains of New Hampshire represent felsic plutons that have been exposed at the surface. Half Dome in Yosemite National Park, California, is an exposed bulge of a granitic pluton.

Granitic rocks have a particular importance for the science of geology because of the accessory minerals they tend to contain as part of their suite of minerals. These accessory minerals

Above: This xenolith from near Larchmont, New York, was at one point a solid within the granitic magma from which the surrounding rock formed. Right: Rhyolite is the fine-grained equivalent of granite but can contain some crystals visible to the naked eye. Top left: A thin section of felsic mylonite, seen under cross-polarized light, shows a quartz crystal in the center with undulatory texture. Undulose quartz has been subjected to tremendous squeezing, and indeed mylonite is a metamorphic rock that has undergone extreme pressure deformation.

Above: View from atop Cadillac Mountain at Acadia National Park in Maine. Jointing (linear fracturing) is visible on the relatively smooth surfaces of this Cadillac Mountain granite. Left: Half Dome in Yosemite National Park, an exhumed batholith (massive body of granitic rock), part of the series of the granite plutons that form the Sierra Nevada in California.

are volumetrically minor in comparison to the feldspar and quartz of a granite, but they can be absolutely essential for

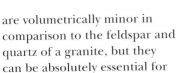

providing a radiometric date for the rock. For example, the mineral zircon ($ZrSiO_4$) is a favorite target for geochronologists because it tends to seal in isotopes of uranium and its daughter product, lead, in such a way that extremely accurate radiometric dates can be derived from the isotope ratios preserved in the zircon crystal.

XENOLITHS

Occasionally, granites will contain chunks or pieces of other types of rocks. Granites in the Sierra Nevada and elsewhere are known to contain crystallized blobs of mafic rock. These blobs, known as mafic inclusions or xenoliths, provide evidence of a variety of mixing processes that involve

both felsic and mafic magmas. Typically, the mafic magmas originate from depths that are greater than the depth of origin of the felsic magmas, and the mafic magmas are necessarily injected upward into the felsic magma body.

In other situations, as for instance Cadillac Mountain of Mount Desert Island, Acadia National Park, Maine, angular chunks of rock that once formed intact layers above the rising felsic pluton were shattered and fell downward into the rising magma. These blocks, now metamorphosed by the heat of the cooling granitic magma, can still be recognized as having originally been derived from rock types such as sandstone, limestone, and the like.

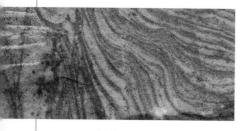

Clastic Sedimentary Rocks

As igneous rocks weather, they break into blocks, chunks, and grains that are collectively referred to as clasts. Over time, these clasts are broken down further into yet smaller particles. The process ends only when the fragments are collected together as loose sediment and lithified or turned to stone in what is called a clastic sedimentary rock.

SEDIMENT GRAIN SIZES

Analogous to the case of igneous rock classification and crystal size, clastic sedimentary rocks are classified primarily by the sizes of clasts that they contain. For example, a conglomerate has clasts ranging from 0.78 inches (2 mm) to 6.5 feet (2 m) in diameter, a sandstone has grains ranging from .063 millimeters to about 2 millimeters, siltstone is composed of particles ranging in size from 4 microns to 63 microns, and mudstone is composed of very tiny particles smaller than 4 microns in diameter. All of these rock types—conglomerates, sandstones, siltstones, and mudstones—are considered clastic sedimentary rocks.

Siltstone, a common type of sedimentary rock that forms in relatively quiet, aquatic environments such as shallow seas, riverbanks, and lakes.

Shale is the most abundant type of rock exposed at the Earth's surface (or under the sea for that matter), and it usually consists of thinly bedded mudstones sometimes alternating with fine siltstones. This sedimentary rock type is so abundant because mud particles consist primarily of tiny crystals of clay minerals. These clay mineral particles are formed chiefly by the weathering of feldspars, and this in part explains why such clay minerals are so widespread. Feldspar is the most abundant type of mineral in igneous rocks.

CLASTIC SEDIMENT MATURITY

In considering clastic sedimentary rocks, the concept of maturity is essential to understand the genesis of these rock

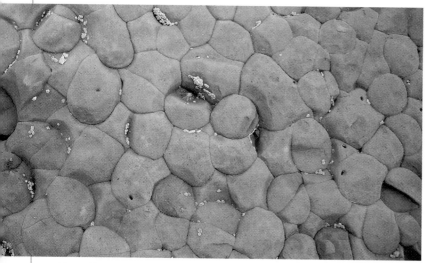

Above: Fractures appearing in this mudstone formation on Lyme Regis East Beach in Great Britain are due to rapid weathering in a shoreline environment. Lyme Regis mudstone is famous for its fossils of Mesozoic marine creatures. Top left: Ferruginous sandstone showing the peculiar liesegang banding, a pattern of parallel bands of iron-oxide-rich layer. It is attributed to salty fluids passing through porous sandstone.

types. In general, a more mature clastic sedimentary rock type is one that has sedimentary particles or clasts that have traveled farther from their source and hence have been subjected to more abrasion and chemical attack than have clasts that have remained closer to their source. Thus, a mature sandstone has more rounded sand grains than does an immature sandstone. The grains in a sandstone are typically quartz grains, but sandstones can in theory be composed of grains of any type of mineral. An immature sandstone will typically contain many grains of feldspar as well as grains of quartz, whereas a very mature sandstone will have no, or virtually no, feldspar remaining. This

is so because feldspar is softer and less resistant to chemical weathering than is quartz, and as the sediments that are going to eventually be lithified into a sandstone are carried farther from their source, the feldspars tend to be destroyed in transit. These feldspars transform en route into clay particles that can then travel even farther from the source and be deposited in the quiet waters of a sedimentary basin as shale.

LITHIFICATION

Once sediments have been deposited, in other words have reached their final resting place and are no longer subjected to the mixing and transporting action of wind and water, the process of lithification can

begin. Fluids moving through the pore spaces between the grains in sediment frequently deposit thin mineral cements between the grains. These cements, often deposited in multiple layers, bind the grains together and are primarily responsible for the process of lithification. A mature quartz sandstone that is cemented by, say, a silica (quartz) cement, can become among the hardest and most chemically resistant of all rock types.

Argillaceous shale is characteristically formed in lake or ocean basins by the settling out from suspension in water of fine particulate sediments such as clay. Its formation is usually associated with slow-moving or stagnant water.

Chemical Sedimentary Rocks

One type of sedimentary rock is not formed from loose sedimentary particles or clasts but instead is formed as chemical precipitates. These are known as chemical sedimentary rocks. They typically form on the bottom of a lake or sea but

in some cases can be formed in the water column and then sink to the bottom as an accumulation of tiny crystals all of the same type.

TRANSPORT IN SOLUTION

Like clastic sedimentary rocks, chemical sedimentary rocks are ultimately the result of the weathering and breakdown of older rocks. As a rock undergoes weathering and disaggregation, any part of it that does not remain as a solid clast is dissolved or otherwise placed

into solution as a solute in what originally starts out as fresh rainwater (unless, of course, it is weathering beneath the ocean surface). A variety of chemical processes can get these ions, once part of a rock, into solution. Because it falls through the atmosphere, absorbing carbon dioxide on the way down, rainwater is naturally slightly acidic because carbon dioxide in solution transforms into carbonic acid. If such a raindrop falls on an outcrop of exposed limestone, the carbonic acid will dissolve the mineral calcite that constitutes the limestone. Calcium from the calcite in the limestone thus goes into solution and is carried down gradients in streams or in groundwater toward the sea in most cases.

One of two things can then happen to these calcium ions. In the first situation, the ions are intercepted and deposited as a cement as the groundwater that contained them passes through originally loose sediment. In this case, the calcium resides as what is essentially a very thin layer of limestone binding the grains (of quartz sand, gravel clasts, or whatever) together.

Below: A thinly bedded sample of micritic limestone. Micrite forms from suspended lime mud that settles to the seafloor. Left: Coquina, a type of limestone formed by weakly cemented shell fragments. The composition of this crumbly rock is the same as that of its constituent shells, namely, the minerals calcite and aragonite. The name is derived from the Spanish for "little shell." Top left: Calcite formations from an underground cavern.

A sample of unbedded limestone formed from lime mud and tightly cemented shell fragments. A coiled shell, probably a gastropod, is visible in the lower half.

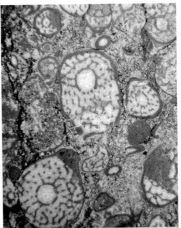

Thin section of archaeocyath limestone showing cross-sections of archaeocyaths and intergrowing calcium carbonate-secreting microbes. Archaeocyaths are spongelike marine organisms.

AQUATIC DEPOSITION

In the second situation, the calcium is carried all the way to a large body of water such as a lake or ocean. Here, the calcium can be combined with the carbonate anion as is the case with the calcite cement mentioned above, but can form as a thick layer of what is in many cases an essentially pure deposit of the mineral calcite to form limestone. Limestones can form in one of two ways. They can form from direct precipitation from seawater, as when seawater supersaturated in calcium carbonate (in solution) wells up into shallow water, as for instance near the shoreline of a tropical beach. Here, the water will warm and no longer be able to hold the calcium carbonate in solution. This will then precipitate as limestone, and sometimes this precipitation occurs in spherical layers around pre-existing grains to form small, concentrically layered balls approximately a millimeter in diameter called ooids. Limestones that form in this way are called oolitic limestones.

BIOMINERALIZATION

Alternatively, organisms such as algae, crinoids, and corals absorb the calcium to form their calcitic (or in some instances, aragonitic) skeletons. Aragonite is similar in composition to calcite but has a different crystal form (needle-shaped as opposed to blocks). These skeletons formed of calcium carbonate can then fall to pieces on the death of the organism, accumulate on the seafloor, and form what is called a bioclastic limestone. Limestones of this nature are often associated with reefs formed by corals, sponges, and other types of shell- or skeleton-forming creatures.

ARCHAEOCYATHAN LIMESTONE

Animals such as corals and sponges are able to secrete massive amounts of calcite during their growth to form reef limestones. The oldest reefs formed by animals were the archaeocyathan limestones of the Cambrian period. Archaeocyaths are thought to be related to modern sponges. Unlike modern sponges, which can have skeletons of calcium carbonate, silica, or tough organic compounds, archaeocyaths had skeletons composed exclusively of calcium carbonate in the form of the mineral calcite. Archaeocyaths underwent a tremendous explosion in diversity during the Early Cambrian, developing a vast array of coral-like forms such as flat cones, clustered colonies, and undulating columnar cones.

Igneous Rocks

Plutonic igneous rocks are those that form beneath the Earth's surface and hence have slow cooling rates and large crystals because they are surrounded by other types of rocks that act as insulation. Volcanic rocks are erupted at the Earth's surface and can spread out to form extensive lava flows. If a volcanic lava is mafic, it will tend to flow easily because its high proportion of iron and magnesium and correspondingly low proportion of silica creates a runny rather than viscous lava for a given temperature. Any gases contained in the lava will try to escape as the lava is erupted at the surface (a relatively low-pressure environment), and in a mafic magma these gases will have a relatively easy time escaping as bubbles that rise and pop at the surface of the hot magma flow.

FELSIC MAGMAS

Felsic magmas, on the other hand, are much richer in silica and are correspondingly much more viscous than a typical mafic magma. This is so because, even when very hot, the silica tetrahedra in a felsic magma are already striving to bond together and tend to slow the flow of the magma. Hence, any gases that try to escape from the felsic magma as it is erupted and

Above: In 1980, a magnitude 5.1 earthquake shook Mount St. Helens. The bulge and surrounding area slid away in a gigantic rockslide and debris avalanche, releasing pressure and triggering a major pumice and ash eruption of the volcano. Thirteen hundred feet (400 meters) of the peak collapsed or blew outward. As a result, 24 square miles (62 square kilometers) of valley was filled by a debris avalanche. Top left: A very fluid volcanic lava flow, Kilauea, Hawaii. When cooled, this lava will form a glassy rock called pahoehoe.

undergoes decompression at the Earth's surface tend to become trapped. This is the fundamental process that leads to a rock known as pumice, formed from a felsic magma and with so many bubbles of trapped gas (forming a kind of glassy froth) that when cool, it can float on water. A variety of marine organisms are thought to have been able to cross ocean basins by attaching themselves to floating pumice.

If enough gas gets trapped in an erupted felsic magma, and if the internal pressures get high enough relative to the atmospheric pressure outside, tremendous explosions can then occur. This is what happened in 1980 during the fatal eruption of Mount St. Helens, when the side of the volcano exploded in a massive release of the pressure built up in its sticky, felsic magma. When a volcanic explosion of this nature occurs, it launches chunks of rock and volcanic glass particles into the air. Larger blocks rain out of the sky as volcanic bombs, whereas the finer particles are carried aloft and eventually settle out as layers of what is called volcanic ash. This is something of a misnomer, because the deposit formed is less an ash (formed by combustion of organic matter) than it is an accumulation of tiny crystals and fragments of volcanic glass.

TAKING TO THE AIR: TEPHRA

Materials shot out and made airborne by a felsic eruption are collectively referred to as

Fossil tree trunks from the Late Triassic tree Araucarioxylon arizonicum *were petrified by silica released by volcanic ash and taken up by pores in the wood.*

In the aftermath of the 2004 eruption of the Mayan Volcano in the Philippines, clouds of tephra cover the landscape, destroying the local plant life. This week-long eruption led to the evacuation of more than 66,000 people.

tephra. Tephra deposits, as volcanic ash, are important in the sedimentary record because ancient events (preserved as bentonite layers) can be recognized in layered rock sequences. In some cases, these bentonite layers can be tied to individually recognized major eruptions.

Truly great felsic eruptions can spread volcanic ash halfway across a continent, bury forests (as in the case of the Triassic petrified forests of Arizona, inundated by water-lain ash), and smother entire seafloor communities in a preserving layer of ashfall.

Metamorphic Rocks

Metamorphic rocks are formed by the alteration of older rocks. They can form from rocks of virtually any type, be they igneous, sedimentary, or other metamorphic. Beautiful, large crystals often characterize metamorphic rocks, and the nature and composition of these crystals are associated with the pressures and temperatures to which the rocks have been subjected. As is the case with igneous and sedimentary rocks, metamorphic rocks can occur in varieties that are distinguished by the sizes of their constituent crystals. As a general rule, the crystal sizes in metamorphic rocks are larger than those in the sedimentary or igneous rocks from which they were formed.

CONTACT METAMORPHISM

Metamorphic rocks are the products of two primary kinds of processes. Rocks that are meta-morphosed because they come into contact with a hot mafic or felsic magma are products of what is called contact meta-morphism. These metamorphic rocks typically form a relatively thin rind or halo around the igneous rock body that intruded into another body of rock. When, for example, a limestone is intruded by an igneous magma, as was the case in the Cambrian limestones at Nahant in Boston Harbor, a greenish zone of contact metamorphism known as a skarn forms around the intruding igneous body. The skarn zone is greenish because of the presence of epidote, a mineral that has elemental constituents coming both from the limestone (calcium) and the magma (iron, aluminum).

REGIONAL METAMORPHISM

Metamorphic rocks can also form as a result of what is called regional metamorphism. In this case, an extensive region, band, or belt of rock undergoes transformation as a result of grand-scale geological events such as emplacement of a granitic pluton, subduction of a slab of oceanic crust, or the plate tectonic collision of two once-separate continents. Grand events of this nature subject rocks nearby to tre-mendous increases in pressure and temperature, and entire bedrock terranes can be transformed as a result. In such situations, metamorphic petrologists can identify these concentric zones of metamor-phism, with the most intense zones of metamorphism occurring closest to the heat or

Above: Green quartzite, a rock formed by the metamorphism of sandstone. Although distorted, the individual sand grains may still be visible in the quartzite when viewed in petrographic thin section. Top left: A massive outcrop of Cambrian age limestone (about 550 million years old) in Nevada's Great Basin National Park forms many caverns, including the Lehman Caves.

pressure disturbances and the least intense zones of metamorphism occurring farthest away from the heat and/or pressure source, respectively. These concentric metamorphic zones are known as metamorphic grades, and are named for the minerals that are most characteristic of the rock that forms as a result of metamorphism at any particular temperature and pressure. For example, high-grade metamorphism will result in rocks belonging to the amphibolite facies, named for the amphibole-bearing rock (amphibolite) that is characteristic of this level of metamorphism. Low-grade metamorphism will result in rocks belonging to the greenschist facies, named for relatively weakly metamorphosed rocks that have a greenish tint due to the presence of the low-grade metamorphic phyllosilicate mineral chlorite.

PARENT ROCK OR PROTOLITH

In the scientific study of metamorphic rocks, a primary challenge is to identify the type of rock that the metamorphic rocks in question were derived from. This precursor rock type is referred to as the parent rock, or protolith. So for instance, a coarsely crystalline marble is likely to have a limestone as its protolith. A micaceous (mica-rich) schist containing beautiful red garnets often has rather drab-looking shale as its protolith. In cases of contact metamorphism, it can be relatively easy to identify

the protolith, because a geologist can simply trace the rock body away from the zone of contact metamorphism and find out what the protolith was like by direct inspection. With regional metamorphism, identification of the protolith can be a much more complex task, for in many cases the entire protolith formation has been metamorphosed and there is no unmetamorphosed protolith remaining. This can be an unfortunate situation for paleontologists (all fossils contained within the protolith are likely to have been destroyed), but determining the protolith can still be an essential step in unraveling the geological history of any particular metamorphic region.

Above: Amphibolite, a high-grade metamorphic rock usually formed by metamorphism of pyroxene-rich gabbro, the coarser-grained equivalent of basalt. Below: Garnetiferous mica schist, a metamorphic rock that consists of equidimensional pink crystals of garnet set in a matrix of crystals of mica. Yellowish limonite stains on the rock surface indicate that it contains a significant amount of iron.

CHAPTER 4

WEATHERING AND SOILS

Left: Every type of soil, from rocky sand to sticky clay, is a unique combination of relief, climate, living matter, parent material, and time. Soil science is the study of these physical, chemical, and biological properties. Top: Peat exploration in the nature conservation area of Ewiges Meer, a moor lake in East Frisia, Germany. Bottom: A biological science technician uses an electronic caliper to measure water depth while studying the soil of a no-till plot. Scientists are studying no-till farming as a potential method for reducing the carbon emission of soil.

The structure and geochemistry of soils are important areas of geological study, with vast implications for agriculture, engineering geology, and environmental management. The study of soils, once dominated by textural and color classifications, is now coming into its own as an essential element in our attempts to understand the interactions between the atmosphere, biosphere, hydrosphere, and lithosphere. Soil, with its capillary water and air pockets, contains elements of all four, and thus forms a vital chemical interface for a variety of geological and biogeochemical processes. One of the first scientists to fully grasp the importance of soil processes was Russian soil scientist V. V. Dokuchaev. Russian and Canadian scientists associated with Dokuchaev Soil Science Institute in Russia are leaders in the study of soils of cooler climates. Release of methane and carbon dioxide from these soils, along with release of similar gases from northern peat lands, is being intensively studied as a way to understand the major controls on contemporary climate change. In this regard and in many others, soils are a factor of global importance, and it is no surprise that Dokuchaev had significant influence on Vladimir Vernadsky as he developed our modern concept of the biosphere.

Rock to Regolith

Rocks are broken down by a variety of processes, including chemical weathering, mechanical weathering, frost wedging, and by meteor impact. The gravel and sediments that result are the starting point for the development of soil. All types of terrestrial environments, from alluvial fans to muddy floodplains, can act as starting points for the initiation of soil formation.

REGOLITH
Regolith is a layer of broken rock, soil, and loose sediment forming a mantle over the surface of a decomposing or otherwise fragmented bedrock mass. Regolith is the most primitive type of soil because it has not yet undergone the geochemical processes that modify pulverized rock and transform it into soil. The full suite of these processes is unique to Earth. Other bodies in the solar system, such as the Moon, develop regolith. The Moon, however, lacks the water circulation of a hydrological cycle, not to mention the catalytic activities of soil microbes, and thus regolith on the Moon never develops beyond the most primitive type of soil, as would be the case on Earth.

In spite of Earth's long history of hydrological circulation and microbial weathering, soils on Earth can on occasion start from scratch. For example, after an episode of glaciation, the glaciers melt and retreat and leave behind piles of unsorted rock debris known as till. This till, consisting as it does of primarily fresh pieces of broken rock in a variety of sizes, provides an essentially pristine substrate for the processes of weathering and soil formation to act upon. The loose rocky material (ejecta) thrown out by a major extraterrestrial impact event would also act as a pristine substrate for soil formation.

CATIONS
One of the first things that happens to regolith or till as it begins to weather to soil is the leaching out of a variety of ions as water interacts with the mineral surfaces in the rock

Left: Regolith soil from granulite rock in a high road cut. Also called mantle rock, it is the layer of loose rock resting on bedrock, constituting the surface of most land. Top left: A 1916 photograph of till found in Wayne County, Michigan. Till is the unsorted material deposited directly by glacial ice and shows no stratification.

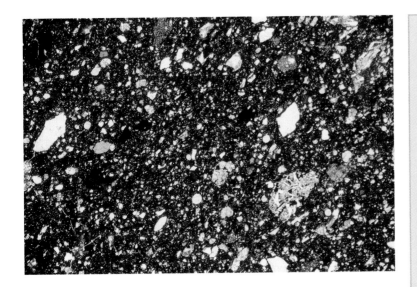

powder. As the feldspars in regolith break down to clays, and organic material begins to accumulate in the soil, we can begin to refer to exchangeable cations, those that adhere to and may be more-or-less easily removed from organic matter and clay particles. Cations (ions with a positive charge) are very important for determining the ultimate fertility of a soil. Sodium, for instance, controls the stickiness and permeability of soil. Too much sodium (it does not take very much to do this) can render a soil too alkaline for optimal plant growth. An excess of hydrogen ion renders soils acidic. Acid soils, such as sometimes form on decomposed granite called grus (pronounced "groos"), can be injurious to plants because the acidity releases toxic aluminum compounds. Better soils have abundant calcium, a cation that helps maintain a beneficial soil chemistry.

ANIONS

Anions (ions with a negative charge) in soil, also known as mineral nutrients, are essential for plant growth. The big four essential anions for plants are nitrate, sulfate, chloride, and phosphate. The mobility of these anions in soil varies. Nitrate, sulfate, and chloride do not adsorb to soil minerals and move unhindered through soil water. Phosphate, on the other hand, clings tightly to clay and other soil minerals, and its mobilization requires the nutrient-absorbing activities of soil fungi.

Top left: A sample of regolith breccia collected by the Apollo 15 *mission, which landed in the Apennine Mountains, part of the ejecta rim of the Moon's Mare Imbrium. Below: Chert breccia, a sedimentary rock consisting of angular, coarse fragments embedded in a fine-grained matrix.*

Soil Science

The study of soils has moved beyond simple classification of soils to an understanding of soils as catalytic chemical factories. Chemical transformations occur in soils that reconfigure minerals and release ions into pore water and into groundwater. Water itself plays a key role here, because its tendency to disassociate into hydrogen cation and hydroxyl anion provides powerful levers for the chemical prying apart of rock-forming minerals in soil. The most basic such reaction is the breakdown of a feldspar (orthoclase) by hydrogen in water. The result of this reaction is freed potassium and aluminum ions, and silica liberated from the feldspar crystal structure. Potassium is a key nutrient for plants and can be easily depleted from soil unless there is a nearby supply that can be replenished through the breakdown of feldspar or other processes. Bicarbonate ions are also present in soil and, like water itself, play an important role in breaking up regolith minerals and freeing the mineral nutrients.

MORPHOLOGICAL STUDY

Studies of the morphology of soil have resulted in fascinating classification schemes that relate soil type to regional climate conditions. There is not necessarily a one-to-one relationship, however, between soil type and present climate regime, because soils take time to mature and climate (particularly during times of glaciation) may change rapidly. There may be a lag time while a soil slowly reestablishes equilibrium with its new climate. For example, a strongly developed soil developed on a stable landscape for 100,000 years and with a mature soil profile (such as, for example, the San Joaquin Soil Series in California's Central Valley) may be out of harmony with the present climate and will begin to change unless it is buried (to become a fossil soil or paleosol) by sediments developing younger soils.

SOIL PROFILES

A soil profile is a narrow cross-section of the land surface that describes the morphology of the

Above: A sample San Joaquin Soil Series. Top left: A substrate of soil will display the varying colors and textures of its component parts.

soil below the surface. Within this soil profile are individual soil units called peds, that might more familiarly be called dirt clods. Peds form in a variety of shapes, from angular to blocky to granular to platy to columnar or prismatic. These peds form the structural units of the soil. Beyond this, soil is described in terms of its texture, color, consistency, and other observable properties such as evidence for root penetration. The sediment content of soil can be plotted on the soil texture ternary diagram. The three corners in this triangular diagram are sand, clay, and silt, respectively. All three sediment size classes can be constituents of soil. A soil that consists of one-third clay, one-third sand, and one-third silt would be classified as a clay loam in the soil texture triangle. The clay content seems to control the overall texture of the soil, and most soils with more than 50 percent clay are considered clay soils. In contrast, a soil needs to be about 85 percent sand before being called a sand soil. Sandy soils, with their good drainage, are better for growing plants than are clay soils, and a very sandy soil can be made better for crops by addition of clay. Another way

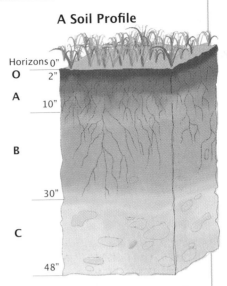

A Soil Profile

Horizons 0"
O   2"
A   10"
B
       30"
C
       48"

A soil profile consists of visually and texturally distinct layers, divided into four parts: O, organic matter; A, the surface layer with most organic matter accumulation and soil life; B, subsoil, where iron, clay, aluminum, and organic compounds accumulate; and C, the substratum of unconsolidated soil parent material.

of saying this is the old farmer's rhyme: Pour sand into clay, throw money away; pour clay into sand, money in your hand.

SOIL CLASSIFICATION

Soil surveys require the use of soil classification. Soil classification is a taxonomic system designed to classify soils in terms of their characteristic materials, diagnostic horizons, and prevailing soil moisture regime. Other schemes have been used to classify soils as well. This classification scheme has great utility and helps to distinguish soils such as entisols (sandy, weakly developed soils) and mollisols (thick, dark soils with a lot of organic matter).

The rhizosphere is the zone that surrounds the roots of plants. As well as roots, it contains other forms of life, such as fungi and nematodes.

ORGANISMS AND SOIL: THE RHIZOSPHERE

One of the textural features in soil identification is the degree of root penetration into the soil. As roots move into soil, they carry with them an active soil community of organisms that makes up what is called the rhizosphere. The rhizosphere includes plant roots, mycorrhizal fungi, and even small animals such as nematodes. These organisms live in symbiosis, and the fungi are the mineral nutrient experts of the community. Fungi are known to use acids in order to bore into feldspar crystals to remove the mineral nutrients.

Caliches and Laterites

The parent material can have a decided influence on the character of soil as it forms, particularly in the early stages of soil formation. As soils age and mature, leaching of ions proceeds. For a given climate regime, soils from different parent materials may tend to become more similar as time goes on. This is particularly so in humid climates where weathering can be intense.

Parent materials can be as diverse as wind and water deposits, glacial debris, and regolith developed on igneous, metamorphic, or sedimentary terranes. But it is really the climate regime that exercises the major control on soil type. For example, in certain desert environments, soils from many types of parent materials will develop a Bk, or calcic horizon. This Bk layer can develop into a caliche layer within the soil. Caliche is a subsurface soil layer hardened by accumulation of calcium carbonate as a pore-space filling precipitate. During the sparse precipitation that falls in arid regions, cations such as calcium are mobilized and carried downward into the soil. But instead of being flushed out into groundwater, the calcium ions are stranded at the Bk layer because the water is absorbed or evaporated before it can flush the system. Caliche layers as hard as concrete have formed in this fashion.

Above: A collection of nitrate caliche in San Bernardino County, California. Top right: Bauxite is most commonly formed in deeply weathered rocks. In some locations, deeply weathered volcanic rocks, usually basalt, form bauxite deposits. Top left: A soil's parent material influences both its color and texture.

SOIL HORIZONS

Under more humid climate regimes, leaching can proceed at such a rate that the most easily mobilized ions are entirely removed from certain layers, leaving them enriched in aluminum and iron oxides, both of which are highly insoluble. Such soil horizons are referred to as the Bo, or oxic horizon. Besides being enriched in aluminum and iron, this soil is depleted in organic matter, which is destroyed by oxidation accompanying the leaching. Oxic horizons are best formed in soils developed in very humid, warm tropical environments. Assuming that the bedrock regolith begins with sufficient iron and aluminum content, a soil type known as oxisol can develop where the soil-forming material has experienced intense weathering over time in a warm, humid climate. Oxisols are also called laterites, a term that refers to soil deposits stained red because of the preponderance of aluminum and especially iron oxides. Most mineral nutrients and organic matter have been leached out of red laterites, therefore these soils do not have what is needed in terms of mineral nutrients for sustained agricultural productivity. Rain forests can nevertheless develop on laterites, where any organic matter that falls to the forest floor is rapidly recycled.

In conclusion, one of the primary ways that climate can influence soil formation is through the rate of leaching or deposition of ions in the soil. Too little precipitation forms caliche, too much forms an oxic horizon or, taken to extremes, a lateritic soil that is in effect an extensive Bo, or oxic horizon. Such soils can be associated with useful mineral deposits such as bauxite (enriched in aluminum oxide) and kaolinite (a light-colored clay deposit formed by intensive weathering of feldspars in a granitic terrane).

Above: The rich, earthy red color of an oxisol is due to the presence of aluminum and iron oxides. Below: A kaolinite mine in China. Kaolinite deposits form by the deep weathering of the mineral feldspar. The finest china is made using clays mined from kaolinite deposits.

Biotic Enhancement of Weathering

The appearance of life on land has been implicated in the acceleration of weathering rates. The rate increased dramatically with the appearance of vascular plants on land, and increased again with the appearance of modern flowering plants.

conclude that this was in fact the case, for moist microhabitats between the particles of sediment in early soils would have been a very favorable habitat for early microbes, particularly bacteria that were adapted to living in fairly thin films of

microbes became the pioneer catalysts of a new geochemical process that David Schwartzman and Tyler Volk have called biotic enhancement of weathering. Regolith influenced by rain alone will slowly undergo weathering. Rain that passes through an atmosphere with carbon dioxide is naturally slightly acid (some of the carbon dioxide dissolves in the rain and forms carbonic acid), and this will by itself act to dissolve and break down mafic minerals, feldspars, and other minerals. The rate is very slow, however, especially for certain refractory minerals. Add

Left: Rendering of microbes. The earliest forms of terrestrial life were no doubt microbial. Below: Even the large crystals of yellow wulfenite are subject to disintegration if microbes multiply over their surfaces. Top left: Moss over a rock. Living organisms from microscopic bacteria to large angiosperms will contribute to the weathering of rock over time.

EARLIEST LIFE

The earliest life on land was no doubt microbial, and although the supporting evidence is slim, many scientists presume that life colonized soils billions of years ago, perhaps not long after life first appeared in the seas (assuming that life had a marine origin). It seems reasonable to

water. Mineral nutrient sources for any such bacteria would have been relatively abundant in these pore fluids because bicarbonate anions (released by the microbes) would act to break down soil minerals and to release their mineral nutrients. Once established on land in the soil and on rock surfaces,

Although most of their Palaeozoic relatives were large trees, staghorn clubmoss (Lycopodium clavatum) is a small evergreen, mosslike plant.

LAND PLANTS

This biotic enhancement of weathering process, however, was carried to a whole new level with the advent of land plants in the Middle Paleozoic, about 400 million years ago. Thanks in large measure to the mineral nutrient mining abilities of their root fungal symbionts, vascular plants such as lycopods and ferns were able to colonize increasingly dry land environments. As their mycorrhizal fungi bored into a previously unexploited nutrient resource (the pristine soil minerals), land plants were able to expand their range into progressively more upland environments. As they did so, the land plants created leaf litter and secreted massive amounts of humic acids. This amounted to a delivery to soil of mineral-attacking acids and other chemicals on a scale that had never before been experienced. Advanced vascular plants such as flowering plants (angiosperms) are even better at doing this than are the more ancient plant groups such as lycopods and ferns. This biotic enhancement of weathering has accelerated both the rate of weathering in general and the rate at which soils age and mature.

microbial life to the setting, however, and the situation changes dramatically. Bacterial waste products such as carbon dioxide can greatly change the pH of soil pore water and destabilize nearby soil minerals. Bacteria and other microbes can actively release compounds that break down soil minerals. The thread-like hyphae of fungi can probe through soil and release acid at the hyphal tips that will actually excavate a tunnel right through feldspars and other crystals. Microbes can multiply over the surfaces of even large crystals of amphibole and other mafic minerals, etching the crystal faces and otherwise causing the crystals to disintegrate.

Four buried paleosols uncovered in the desert in Bernalillo County, New Mexico. Geologists study paleosols to determine a time frame for Earth's oxygenation.

PALEOSOLS AND THE OXYGENATION OF SOIL AIR

Paleosols, or ancient buried soils, have been used to help determine the moment at which the atmosphere began to accumulate significant amounts of free oxygen. Early red-colored soils appear at approximately 2.7 billion years ago, suggesting that oxygenation of the atmosphere was well under way by that time.

CHAPTER 5

A LIVING PLANET

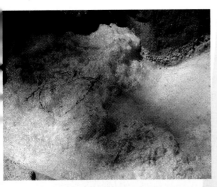

Left: Warm sea conditions promote the growth of anthozoans, coralline algae, and a colorful tropical marine snail. Top: Cyanobacteria colonizing the sediment-water interface. Bottom: A low-temperature sulfide chimney vent on the seafloor, nicknamed "Zooarium," hosts a flourishing colony of tube worms.

Carbon plays an essential role in determining Earth's surface conditions. Both its reduced form (methane) and oxidized form (carbon dioxide) control climate, and both forms (as well as a host of other carbon compounds) are consumed and excreted by a wide variety of organisms. Although they do not constitute the only control on the amount of carbon dioxide in the atmosphere and hydrosphere, organisms play a key role in modulating the concentrations of this gas and are thus thought to play key roles in regulating Earth's climate. For example, warm sea conditions encourage the growth of calcium carbonate—secreting marine animals such as corals. As these corals flourish, they absorb large amounts of carbon dioxide (to make up the carbon in the carbonate), reducing the amounts of carbon dioxide in the hydrosphere (and by extension, the atmosphere) accordingly. Similarly, warm and moist land conditions promote the expansion of land biota, growth of which requires huge amounts of atmospheric carbon dioxide. The biota of both land and sea therefore has a stabilizing influence on climate; as conditions warm, organisms absorb more carbon dioxide and thus help prevent the global warming induced by the greenhouse gases from getting out of control.

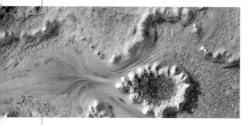

Microbe Earth

Many microbes, particularly bacteria, are talented chemists and are able to oxidize and reduce many types of carbon- and nitrogen-containing compounds in the environment. This microbial ability changes the chemical composition of sediments and the rocks that are formed from them. Understanding of these microbial abilities is essential for understanding the biogeochemistry of planet Earth.

PASTEUR'S INSPIRATION

In 1866 and 1876, respectively, the great French scientist Louis Pasteur (1822–95) published his pioneering studies on fermentation in wine and beer. In one of the greatest East-West scientific exchanges of all time, Russian scientists picked up from where Pasteur left off and extended his work to the geochemical realm. The Russian microbiologist Sergei Nikolaevitch Vinogradsky (1856–1953) discovered the capacity that soil microbes have for chemosynthesis. Vinogradsky's 1887 discovery of the chemoautotrophic abilities of microbes is of overwhelming importance for geochemical studies. Chemoautotrophy is the ability, in the absence of sunlight, to create one's own food from simple gases dissolved in water, such as hydrogen gas, hydrogen sulfide, and carbon dioxide. Russian geochemist Vladimir Vernadsky (1863–1945) drew inspiration from both Pasteur's and Vinogradsky's discoveries and realized that the fermentative, or medium-forming, activities of organisms that Pasteur characterized in alcoholic beverages and cheese applied to broader Earth environments as well.

ATMOSPHERIC INFLUENCE

In the 1960s, American paleontologist Preston Cloud (1912–91) realized that microbial activity had profoundly altered the composition of the atmosphere early in Earth's history. Early types of microbial photosynthesis did not release oxygen; however, a more advanced type of photosynthesis called photosystem II (PSII) used water and carbon dioxide as its raw materials and created oxygen as a waste product. This biochemical process, requiring only common environmental chemicals and sunlight, became a rapid success and spread to many aquatic environments on the early Earth. At first, accumulation of the waste oxygen was slow because there were many unoxidized (reduced) compounds in air and water

Above: A photograph of famous scientist Louis Pasteur. Top left: Filamentous bacteria and algae found in Yellowstone's Upper Geyser Basin. This bacteria serves as a sulfur-oxidizing bacteria.

that could combine with and neutralize the chemically aggressive oxygen. Yet, as oxygen became more widely distributed as a result of the success of the cyanobacteria, a particularly well-adapted kind of bacterium that uses PSII, abundant oxygen began to make its presence evident in the geochemistry of the planet. Marine sedimentary rocks deposited between 2.5 and 1.8 billion years ago known as banded iron formations are evidence of a gigantic oceanic titration event, in which oxygen released by the cyanobacteria combined with oxygen dissolved in seawater to form iron oxide, or rust. This red iron oxide sank to the seafloor in rhythmic pulses, forming the characteristic banded deposits that are such an important modern source for iron ore. Cyanobacteria continue to influence atmospheric composition by their absorption and release of carbon dioxide, as is the case for other photosynthetic organisms such as seaweeds (a type of protist) and land-dwelling vascular plants.

Banded iron formations, such as the one seen here, are among the oldest rocks on Earth. Tectonic movement has caused the rock to deform over time.

REDOX VIRTUOSOS

On the shallow seafloor in many bays, harbors, and lagoons, the sediments are covered with a yellowish fuzz of sulfur bacteria. These bacteria, called *Thioploca*, and associated microbes are responsible for both sulfate reduction and methanogenesis (the biosynthesis of the hydrocarbon methane) in marine sediment. These biogeochemical processes involve the concept of redox, or complementary oxidation and reduction. In oxidation, electrons are taken away from an electron donor molecule and transferred to an oxidizing molecule. In reduction, electrons are added to an electron receptor molecule (thus reducing its oxidation state) and taken from a reducing molecule that provides the electrons. Both processes, oxidation and reduction, can be used by microbes to generate biochemical energy (stored as sugars). The process to use depends on the oxidation state of the surrounding environment, namely, whether the local environment is oxidizing or reducing in itself. For example, in many areas, a few centimeters below the sediment-water interface on the seafloor, free oxygen is absent. The sediments are saturated with stagnant or slow-moving seawater, however, and this seawater is loaded with dissolved sulfate. Sulfate-reducing microbes strip oxygen off the sulfate in this environment, using it to metabolize just as humans use oxygen in the air.

Methane and Gas Hydrates

Clathrates, or gas hydrates, are a bizarre form of ice that contains gas molecules such as methane in molecular cages. Hydrate layers are extremely widespread in subsurface layers in the oceans and on land in Arctic regions. They play a potentially very important role in regulating the global carbon cycle. For example, gas hydrates decomposing on the continental shelf edge and in the cold seeps of Monterey Canyon off the coast of California release methane into the sea, supporting communities of tube worms and other organisms that feed on the seeping methane.

GAS HYDRATE CHEMISTRY

When the first sediment cores containing gas hydrates were brought on board the decks of ships that were being used as drilling platforms, the cores tended to explode because hydrates break down at room temperature. Stable only at the elevated pressures and cool temperatures found beneath the seafloor, gas hydrates have a chemical structure that consists of gas molecules such as methane trapped inside a polyhedral cage formed by linked water molecules. The clathrate structure is a regular one, as is any crystalline solid, and it is curious how a low-molecular-weight gas such as methane can participate with water in the construction of this strange icelike solid. Other hydrocarbon gases of higher molecular weight, such as ethane, can also become trapped within the clathrate structure.

GEOLOGICAL OCCURRENCE

Gas hydrates are extremely widespread within seafloor sediments. Although generally not visible on the seafloor itself, layers of gas hydrates show up prominently in seismic profiles of the seafloor made by ships towing hydrophones. The gas hydrates appear as bottom-simulating reflectors, or BSRs. The shape of these BSRs reflect the topography of the seafloor above them; for example, where the seafloor rises up toward the surface, as on the Blake

Above: Methane seep with gastropods sustained by bacteria that have colonized the seep area. Top left: Computer-generated three-dimensional image of the seafloor of Monterey Canyon. Cold seeps within the canyon support communities of methane-feeding organisms.

carbon gas components. Were this to occur on a massive scale, the long-term climate implications are uncertain but quite possibly severe. Methane is approximately 20 times more potent as a greenhouse gas than is carbon dioxide.

HYDROCARBONS IN THE HYDRATE RESERVOIR

Considering both its marine and land occurrences, the global reservoir of methane gas hydrate and other hydrocarbons exceeds the amount of all our remaining oil and gas deposits. This huge amount of hydrates, amounting to some 10,000 gigatons, could play a disastrous role, altering climate were it to be all released suddenly. On the other hand, were it possible to extract and use these hydrocarbon gases, an abundant new energy resource could suddenly become available. Unfortunately, to date no one has discovered a cost-effective means of recovering the gas from gas hydrates. It is theoretically possible to do so, but one must be able to simultaneously break down the hydrate and collect the gas in an economically feasible way. Unlike free natural gas and oil, gas hydrate tends to clog the pore spaces between sediment grains, and that means you have to break down the hydrate structure both in order to free the gas and to keep it flowing.

This core sample, taken from the Mackenzie Delta, contains visible white gas hydrate crystals. Located in the Canadian Arctic, the Mackenzie Delta has an extremely high concentration of gas hydrates.

Bahamas Ridge, the hydrate layer rises toward the surface as well. This is so because the pressure and temperature zone that favors gas hydrate formation generally occurs at a particular depth below the seafloor. This depth is variable, and depends partly on the temperature of the ocean in any particular area.

An equally interesting reservoir of gas hydrates occurs anywhere in the world where permafrost forms. Most of these regions tend to be in the Northern Hemisphere, and vast gas hydrate deposits occur in Siberia, Alaska, and Arctic Canada. As global climate warms, these land hydrates will melt and release their hydro-

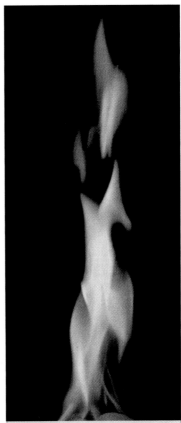

Burning snowball sustained by methane released by a ball of gas hydrate ice.

FLAMING SNOWBALLS

Flaming snowballs are formed of gas hydrates. The fire is due to combustion of hydrocarbon gases that are released as the gas hydrate melts at room temperature. The organized molecular structure of clathrates breaks down if not kept cool and under pressure. Icelike gas hydrates trapped within subsurface sediments may hold promise for abundant new energy resources. A major challenge is getting the gas out of the hydrate in economical quantities.

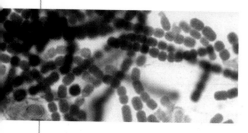

The Carbon Cycle

Coral reef limestone, oil and coal deposits, and atmospheric carbon dioxide are all examples of global reservoirs of carbon. Carbon compounds both regulate the temperature of the atmosphere and control the acidity (via carbonic acid buffering) of seawater. Both are essential for the continuation of life in the seas and on land. Changes in atmospheric carbon can put the world into greenhouse or icehouse conditions.

SECULAR DROP IN ATMOSPHERIC CARBON DIOXIDE

Several lines of evidence suggest that, early in Earth's history, the carbon dioxide content of the atmosphere was higher than it is now, and that (at least from the perspective of modern humans) the climate has gotten better with time. This is not just a function of the fact that we are relative newcomers to the biosphere and evolved and adapted in accordance with the currently prevailing climate. Very high atmospheric temperatures have a deleterious effect on the survival of complex multicellular organisms such

Below: Diagram of the carbon cycle, showing how carbon moves between the atmosphere, hydrosphere, and geosphere. The annual carbon fluxes are in red and the total amounts of stored or sequestered carbon in black. Top left: Cyanobacteria, distinctive for their blue-green hue.

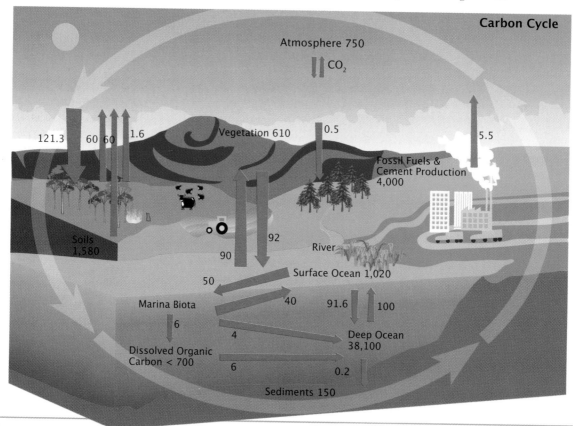

Carbon Cycle

Atmosphere 750

CO₂

121.3   60 60   1.6   Vegetation 610   0.5   5.5

Fossil Fuels & Cement Production 4,000

Soils 1,580

92

River

90

Surface Ocean 1,020

50

Marina Biota

40   91.6   100

6

4

Dissolved Organic Carbon < 700

Deep Ocean 38,100

6   0.2

Sediments 150

as eukaryotes, and convincing arguments can be made that complex organisms could not evolve at all until Earth's surface temperatures passed below a certain threshold temperature.

When organisms such as cyanobacteria create organic matter by photosynthesis, they release oxygen into their environment and absorb carbon dioxide. This can tend to reduce the global budget of atmospheric carbon dioxide in two primary ways. First, most of the organic matter in the form of hydrocarbons created by the cyanobacteria will later be oxidized (combined with oxygen) and released back into the environment as carbon dioxide; however, since these microbes live close to or even in sediments, some of the hydrocarbon matter they produce will escape the recycling process and be sequestered in the sediments. When this happens, there is a net gain of oxygen to the atmosphere and a net loss of carbon dioxide to the atmosphere. This imbalance will persist as long as the sediments remain buried or (if lithified) remain as rock. Second, organic acids produced by the microbes will promote the Urey reaction, a mineral-weathering process that absorbs carbon dioxide from water and air. These two factors, carbon burial and rock weathering, worked in concert to cool the planet and to save Earth from the fate of Venus, where greenhouse gases render the planet's surface uninhabitable.

The metabolism of Methanobacteriaceae *results in the production of methane.*

THE PALEOCENE-EOCENE THERMAL MAXIMUM
The Paleocene-Eocene Thermal Maximum (55 million years ago) is apparently associated with an astonishing increase in biogenic, atmospheric methane. Several unanswered questions are associated with this methane increase. Was the archeabacterium *Methanoculleus submarinus* the source of this methane? Was thawing of gas hydrates and methane release the main source? Answering these questions would go a long way toward reducing uncertainties about the impact of methane and other greenhouse gases on future climate.

KOENE'S INSIGHT
In 1856, Belgian scientist Corneille Jean Koene (died c. 1862), an associate of the great Swedish chemist Jons Jakob Berzelius (1779–1848), became the first to realize that carbon dioxide levels on Earth had been falling steadily over geologic time. This was so because photosynthetic organisms, and particularly vascular land plants, utilized increasing amounts of carbon dioxide from the atmosphere. Koene also realized that the concept of atmospheric pollution is a relative term, and that many natural phenomena can cause significant change to the atmosphere. These changes may be good for some organisms and bad for others, but the main message is that the Earth's atmosphere has been in a state of constant flux since its inception 4.5 billion years ago. Vernadsky's concept of the biosphere incorporates Koene's insight, and Vernadsky to his credit emphasized the pressure of life as well as its Pasteurian, medium-forming tendencies. For Vernadsky, the pressure of life represented life's impetus to exploit any resource, colonize any environment it could, almost like the molecules of a gas rushing into a vacuum. In some circumstances, the pressure can get too high and the environmental vessel can shatter. When this occurs (as for instance during the oxygenation of the atmosphere over two billion years ago), a new geochemical regime is put into place on Earth.

The Conquest of Land

The advent of land plants in the middle Paleozoic was such a success in terms of creating biomatter that it led to massive generation of an essentially new rock type, coal. The generation of coal in the aptly named Carboniferous period drew so much carbon out of the atmosphere that it triggered an ice age. The expansion of Hypersea (land plants, soil fungi, and their symbionts) continued into the Mesozoic (gingko mangroves) and continues to this day. A process known as hypermarine upwelling brings mineral nutrients from rocks up to the soil surface and provides a tremendous nutrient supply.

ORIGINS OF FUNGI

Complex life is sometimes divided into four kingdoms: animals, plants, protists, and fungi. Fungi probably originated in the sea, but a minority opinion posits that they first evolved on land, and, unfortunately, the early fungal fossil record is sparse enough to not be able to settle the question. In

Right: A plate from Ernst Haeckle's Kuntsformen der Natur, *depicting a variety of plants, all with fungal symbionts. Top left: Moss and associated microbes covering an otherwise bare rock surface.*

any case, fungi did not become a prominent part of the Earth's biota until they initiated their partnership with the earliest vascular plants. Vascular plants are those, such as mosses and flowering plants, that have internal water-conduction tubes (usually called tracheids) that transport water (via transpiration) from soil-pore water.

HYPERMARINE UPWELLING

When the fungal-vascular plant partnership became established, probably sometime in the Silurian, massive amounts of crustal nutrients were airlifted by the stems of vascular plants up to where the early leaves could really make use of them in full sunlight. This led to the most massive expansion of biosphere since the origin of life in the first place. The expansion was made possible by the osmotrophic abilities of fungi, osmotrophy being the ability to chemically scavenge food such as mineral nutrients from a variety of media. The media (soil particulates, mineral crystals) are changed as a result, and thus this ability of fungi represents a major element of the medium-forming ability of life on Earth emphasized by Vernadsky. The potent combination of mineral nutrients, sufficient water, and

An unpolished hand sample of the Rhynie chert from Rhynie, Scotland. Stems of fossil plants are visible in longitudinal and cross-section.

THE RHYNIE CHERT AND THE MYCORRHIZA-VASCULAR PLANT CONNECTION

The fungal genera *Glomus* and *Sclerocystis* represent evolutionarily ancient types of fungi that evidently were the first to initiate the mycorrhizal symbiosis with vascular plants. Fossils of both types of fungi have been found in the Early Devonian Rhynie chert, a deposit that contains the oldest-known relatively intact fossil plant wetland community. The tiny quartz crystals of the chert are responsible for the exquisite preservation of the plant fossils, many of which occur with fungus preserved inside of them.

Chunks of anthracite coal, a valuable energy resource.

The Conquest of Land

(continued)

sunlight led to the utilization of atmospheric carbon dioxide by living forms on an unprecedented scale. The resultant drop in the atmospheric concentrations of this greenhouse gas may have thrown the world into a major ice age late in the Paleozoic.

COAL AND GLACIATION

The Late Paleozoic was marked by a major glaciation between about 355 and 255 million years ago. The leading hypothesis for the cause of this well-documented glaciation is the expansion of the land plants and their fungal partners as the land biota really began to overcome nutrient limitations and to generate large amounts of organic matter. The plant-fungal partnership created such enormous quantities of organic matter, such as leaf litter, that bacteria and saprotrophic (decay-inducing) fungi could not keep up in terms of decomposition. The organic matter accumulated in sedimentary basins, first to form peat, and later to form coal. Great amounts of carbon dioxide were removed from the atmosphere in this fashion, and it was this drawdown of the carbon-dioxide greenhouse gas that precipitated

the Late Paleozoic glaciation. There is some dispute about this hypothesis for the cause of the glaciation, however. For example, geologist M. R. Saltzman has argued that this glaciation was due in whole or in part to a tectonic closure of a near equatorial seaway during the assembly of the great supercontinent, Pangea. Basing his conclusions on carbon isotopes in the Carboniferous sequence of strata at Nevada's Arrow Canyon in the United States, Saltzman argues that closure of the seaway altered climate by enhancing the poleward transport of warm, moist air that, when cooled at higher latitudes, caused the snow and ice accumulations that led to glaciation. Saltzman's arguments notwithstanding, it would seem that the Earth of the Carboniferous was predisposed to falling into a glaciation because of the plant-induced pressure on the global atmospheric carbon budget.

Right: This 1868 rendering of the Carboniferous (340–280 million years ago) landscape is still considered to be a fairly accurate reconstruction. Mangrove-rooted Cordaitales *and* Lepidodendron *trees are shown in this early forest. Top left: The mushroom-forming fungus* Mycena leaiana.

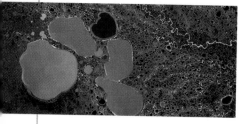

Climate Change

The climate of the Earth, as well as the frequency and intensity of atmospheric phenomena such as storms, seems to largely depend on fluctuations among the main reservoirs in the global carbon cycle. All life-forms play some role in influencing these reservoirs, and our species is no exception to this rule, although the true magnitude of our impact on climate remains to be seen. Might it be possible to imagine ways for us to control climate by managing the global carbon budget?

SIBERIA THAWING

In these times of global warming, Siberia seems to be warming faster than anyplace else on Earth. The Siberian permafrost fields are losing their lining of gas hydrates and are being transformed into a Minnesota-like region of thousands of lakes. The process seems to be accelerating. Sergei Kirpotin of Tomsk State University in Russia fears, with the entire western Siberian region undergoing meltdown, that an irreversible ecological landslide is already under way. It may be premature to panic about this development, however, for it is important to recall that the Siberian permafrost peat fields are only about 11,000 years old and have successively formed, melted, and formed again over the past several million years.

Left: A replanting and reforestation project taking place in Tolotama, Burkina Faso. Top left: Satellite image of Siberian pothole lakes, which were formed by the thawing of permafrost that lost its gas-hydrate lining.

WET AND DRY PEAT BOGS

Peat bogs, be they freshly emergent from permafrost in Alaska or Siberia, or from warmer areas such as Mer Bleue in Ottawa, Canada, are a topic of particular concern because of their potential for generating large quantities of the potent greenhouse gas methane. According to Jill Bubier, a scientist studying Mer Bleue at Mount Holyoke College, if bogs dry out during times of climate change, their potential for release of dangerous methane is limited because the methane tends to oxidize before it reaches the atmosphere. On the other hand, if warming occurs while the bogs are saturated with water, methane can be released directly into the atmosphere. Methane's residence time in the atmosphere is much less than that of the carbon dioxide produced by drying bogs, but it is so much more powerful (by a factor of 20) as a greenhouse gas that a steady supply of methane to the atmosphere could compensate for the oxidation losses and enhance the potential for dramatic short-term climate change.

CAPTURING THE HYDRATES?

From our perspective as humans, a radical shift to a new atmospheric-geochemical regime might not be a particularly welcome occurrence. Our environmental tolerances are greater than that of any other complex type of organism, yet nevertheless even we have our limits, and certainly it is not too hard to displace us from our comfort zone. Evidently world climate is currently in a precarious and unstable state, and we can only hope that there are enough negative feedback systems in the global carbon cycle to forestall massive disruptions to global climate. Certainly there are things our species can do to help tilt the odds in our favor. Reforestation and sustainable forestry are particularly attractive methods for absorbing excess atmospheric carbon dioxide. Fertilizing the seas with particulate iron and silica to enhance the growth of phytoplankton has been tested as a way to soak up carbon dioxide. The world awaits, however, the clever scientist who can show us how to capture the methane as it escapes from the thawing gas hydrate reservoirs, thus simultaneously protecting the atmosphere from this potent greenhouse gas and providing the world with a huge supply of relatively clean-burning fuel.

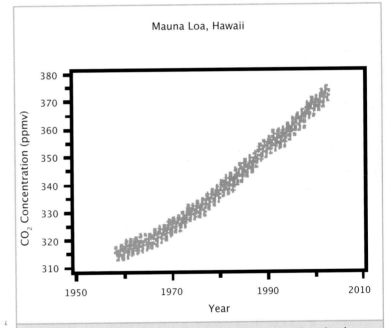

The Keeling curve shows a steadily increasing concentration of carbon dioxide in our atmosphere ever since measurements began in 1958.

ATMOSPHERIC CARBON DIOXIDE LEVELS 1958-2007

Atmospheric carbon dioxide levels have been increasing in a steady fashion ever since precision measurements were first kept at Mauna Loa Observatory, Hawaii. This is the longest record of its type in the world, and the equipment and methods used to gather the data have been essentially unchanged for nearly half a century.

CHAPTER 6

THE FOSSIL RECORD

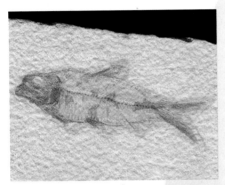

Left: A gigantic fossilized ammonite shell from Mesozoic strata in the Caucasus Mountains. Top: Fossilized remains of Knightia, *a Cenozoic freshwater fish from the Green River Formation. Bottom:* Asterophyllites *fossil sphenophyte plant from the Carboniferous period.*

Earth's sedimentary rocks are laced with traces of ancient life-forms. These fossils provide evidence for the evolutionary sequence of changes in organisms though time and are also a means of dating the rocks. The fossils occur in three types: body, trace, and chemical fossils. Body fossils are generally the most obvious and usually consist of shell, teeth, bones, wood fragments, and other hard parts of once-living organisms. Typically, body fossils consist of either biomineralized material, or resistant organic material such as cellulose or lignin, or tough organic compounds such as sporopollenin. Trace fossils represent disturbances in loose sediment (or sometimes boring into hard rock) by the motions of animals and other organisms. Dinosaur trackways and worm burrows are examples of trace, or ichnofossils. Because trace fossils are part of the fabric of the rock, they are among the most durable and are the last fossils to disappear as rocks are metamorphosed. Chemical fossils are identifiable chemical remains of organisms, and are also known as syngenetic biomarkers. The term *syngenetic* refers to the fact that the chemicals formed at the time the life-form was still alive. Such biomarkers can remain intact in sedimentary rocks for billions of years.

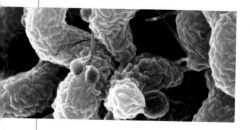

Principle of Faunal Succession

One of the most significant discoveries in the geological sciences is the recognition that the composition of Earth's biota has changed dramatically over time. In general, the more complex the organism, the faster and more complete the biotic turnover for any given interval of geologic time. For example, there are bacteria alive today that are probably very similar to their ancestors of billions of years ago. On the other hand, not a single mammal species alive today was around, say, 50 million years ago. Since then, there has been a complete turnover at the species level in the composition of the mammalian fauna.

MASS EXTINCTIONS

The reason for this turnover is that species are constantly going extinct and new ones appearing throughout geologic time. The rates of extinction and origination, however, are not constant. For example, the rate of extinction during "ordinary" times is relatively constant, although some paleontologists have argued that it has declined slightly over the past half billion years or so. In contrast, there exist short-lived intervals in geologic history when the rate of extinction skyrocketed, and these times are known as episodes of mass extinction. Mass extinctions are important for delineating major boundaries in the geologic time scale, and are quite practical as well. They represent times of wholesale turnover in the composition of the biota. For example, the animals of the last period of the Mesozoic era (Cretaceous) are for the most part very different from the animals of the first epoch of the Cenozoic or Tertiary (Paleocene). The dinosaurs vanish and diverse groups of new mammals appear in their wake. This pattern is even more characteristic of the boundary between the Paleozoic and Mesozoic eras, associated with the Permo-Triassic mass extinction. At this point in geologic time, there is a nearly complete overturn in the composition of the marine fauna.

Left: Fossilized ammonite shell from the Mesozoic, sectioned and polished to reveal sediment filling some (but not all) of the cephalopod's chambers. Chambers without sediment infill are filled or partly filled by calcite crystals. Top left: A colorized scanning electron micrograph of Campylobacter. *Many bacteria have changed very little over geologic time.*

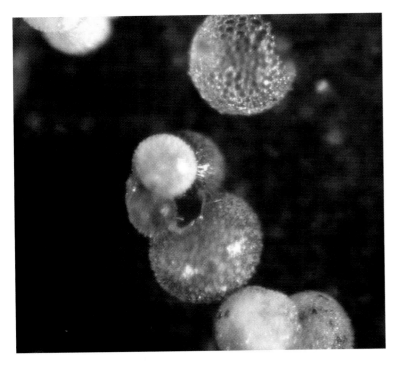

restricted to relatively narrow slices of geologic time. Index fossils should be abundant so that there is a reasonably good chance of finding specimens in the strata whose age is in question. Microfossils such a foraminifera are useful in this regard—they can occur as thousands of specimens per square centimeter of rock. The best index fossils are found in a variety of different sedimentary environments; this allows direct correlation between, say, a limestone and a sandstone. Unfortunately, perfect index fossils are rare because relatively few organisms are both environmentally wide ranging and temporally restricted, probably because the organisms that are more environmentally tolerant tend to survive longer stretches of geologic time.

RELATIVE DATING

The uniqueness of any given biota, particular to any given interval of geologic time, provides geologists with a precise relative dating tool that is used to correlate layers of strata that were deposited at the same time. A sandstone in England and another in France can be shown to be of approximately the same age because they contain identical species of fossils. The technique has now been applied for about two centuries to both marine and terrestrial rock bodies. Marine strata can even be correlated indirectly to nonmarine strata once a sufficiently well-organized list of fossil ranges is available, and this is now certainly the case. In some cases, the comparison can be made directly, as for instance in the case where a dinosaur carcass floated to sea and was preserved in marine strata deposited off the coast of what is now New Jersey.

INDEX FOSSILS

The best type of fossils, however, for these kinds of correlations between strata are known as index fossils. Such fossils are characterized by being abundant, widespread, and ideally

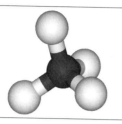

Chemical Fossils

The study of chemical fossils embraces a wide variety of tools and techniques for detecting evidence of early life-forms. This can range from organic 3.8 billion years in age show possible carbon isotope evidence for bacterial life forms. At 3.5 billion years ago, there seems to be biogeochemical evidence for a coincidence that the oldest stromatolites are about 3.5 billion years old.

Above: A stromatolite specimen shown in cross-section. Stromatolites are layered sedimentary structures formed by microbial mat accretion, and are known from rocks as old as 3.5 billion years. Top left: A diagram of a methane atom.

molecules that are essentially unchanged from their living form to inferences about carbon isotopic fractionation that can provide evidence for photosynthesis or other chemical changes associated with living metabolism. Archean rocks ancient sulfate reduction—in other words, where ancient bacteria used the oxygen contained in sulfate to metabolize organic compounds in much the same way that living bacteria do today in low-oxygen environments. It may not be

MICROBES AND METHANE

Geochemical evidence for methanotrophy and methanogenesis appears at 2.8 billion years ago. These carbon isotopes suggest the presence of ancient bacteria that either ate methane (methanotrophs) or generated methane as a waste by-product (methanogens). These isotopes would also be evidence for a beneficial ecological exchange in which the waste products for one type of bacterium become the food sources for another bacteria, and vice versa.

STERANES

An important group of chemical fossil is the steranes. These compounds, which are derivatives of a group of organic molecules to which cholesterol belongs, are characterized by a distinctive pattern of linked five- and six-carbon rings. They are clearly biological in origin, and biogeochemists accept certain types of steranes as syngenetic biomarkers that provide evidence that eukaryotes were present at the time that the rocks that contain them were deposited.

Eukaryotes are organisms with complex cells that contain

organelles and a nucleus, and are considered to be a significant advance over the lowly bacteria from which they are descended. The sterane biomarker record extends back to 2.7 billion years ago, indicating (if correctly interpreted) that eukaryotes, although not quite as ancient as the bacteria, have a fossil record that stretches far back into deep geologic time. Undoubted eukaryote fossils, however, do not appear in the fossil record until about 1.8 billion years ago, indicating that the chemical fossils for any particular group may precede the appearance of their body fossils.

Hymenopteran insects belonging to the same species fossilized in amber. They represent part of an ancient ant swarm.

TAR AND AMBER

The oldest biomarkers that are certainly syngenetic are U-shaped carbon compounds (with side branches) that occur in rocks of the Middle Proterozoic eon, about 1.6 billion years old. By the Late Proterozoic, biomarkers that can be linked with confidence to animals such as sponges can be identified in the rock record. In fossils of the Mesozoic and Cenozoic eras, intact proteins can be identified in well-preserved specimens. Mammal bones from the La Brea Tar Pits, such as those derived from the saber-toothed cat *Smilodon californicus*, have

well-preserved proteins and DNA that were embalmed by the tar and hence protected from bacterial degradation. A great potential source of chemical fossils are in specimens of amber. Amber itself is a chemical fossil you can hold, being derived from the sesquiterpene constituents of resin used by tropical trees to defend against invading insect and fungal pests. These fungi and animals, and other unlucky passersby, get trapped and fixed in the resin and are preserved, along with all their biochemicals, in the amber. In what is currently the ultimate success of chemical paleontology, bacteria have been claimed to have been cultured from the bodies of insects preserved in amber from tens of millions of years ago. Much work remains to be done in the identification and characterization of syngenetic biomarkers.

Fossilized saber-toothed cat Smilodon californicus, *a Pleistocene predator from North America.*

Bacterial Fossils

Body fossils of bacteria are evidently present in rocks older than three billion years of age, but the actual locality of the oldest fossil bacteria is a point of scientific contention and controversy, and no one has yet claimed title to the oldest-known uncontested bacterial body fossil. Part of the problem with reading this part of the fossil record is that, because it is known that bacteria can live quite snugly inside of rocks, it is difficult to be certain that a

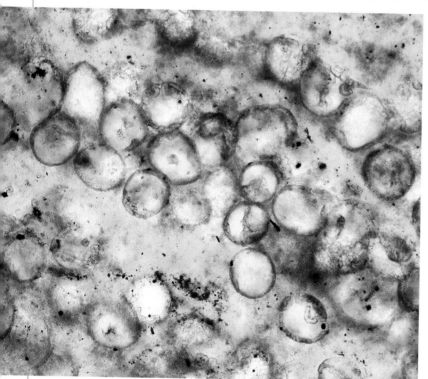

bacterial fossil formed within a certain rock. The potential for scientific conflict is heightened by the fact that a great deal of scientific interest focuses on the first of anything, and therefore

Above: A sampler on the Pisces V *research vessel suctions up an orange microbial mat sample, while an associated probe at lower left records temperature at this underwater volcanic vent. Left: Photomicrograph of the fossil freshwater cyanobacterium* Chlorellopsis coloniata. *These fossils are from the Eocene (55–38 million years ago) of Wyoming. Top left: An aquatic view of a stromatolite from Cuatro Cienegas, Coahuila, Mexico. The syringe is being used to take a sample for chemical analysis; Cuatro Cienegas is being used as a modern analog for the Proterozoic ocean.*

the oldest body fossils are certain to be a focal point for interest, discussion, and debate.

MICROBIAL MATS

Fossil bacteria that have been assigned ages in excess of three billion years mostly resemble modern cyanobacteria, and indeed this is probably the type of organism one might expect to be common at this early stage in the history of life. When cyanobacteria become abundant in any given aquatic environment, they tend to form fabric or feltlike films on the seafloor or lake bottom. These films are known as microbial mats, or biomats, and are geologically important because they can greatly influence depositional conditions for both chemical and fine clastic sedimentary rocks in any given aquatic environment. Filamentous bacteria in particular can form tangled meshworks that both trap sediment particles and precipitate minerals such as calcite.

STROMATOLITES

Where biomats are well developed, they can lead to the construction of mound-shaped structures known as stromatolites. At 3.5 billion years in age, the oldest stromatolites apparently represent the oldest megascopic (that is, visible to the naked eye) fossils in the fossil record. As is the case with bacterial body fossils, the oldest stromatolites have been a subject of scientific controversy. At present, however, the weight of scientific opinion seems to

Patches of encrusting green and red algae overlap communities of filamentous bacterial mats on rock surfaces on the East Diamante volcano. We see here overlap between the chemosynthetic and photosynthetic communities.

favor the view that the most ancient stromatolites (from rocks in Australia) are indeed biogenic and hence the oldest direct evidence for life on Earth. A telling piece of inferential evidence is the presence of different stromatolite forms representative of different local depositional environments. This suggests that the interaction between microbial mats and sediment deposition will change in character from place to place at different points in a shallow nearshore marine setting.

SPACE FOSSILS?

The controversy over recognition of the earliest fossils on Earth is of great import for the science of exobiology—in other words, the search for life beyond the Earth. Many of the same criteria and arguments used to evaluate the earliest Earth fossils and putative fossils will be, and in fact are being, applied to extraterrestrial environments such as Mars. Some of the fossil stromatolites on Earth from the Proterozoic are the size of hills, and such things on other planets could potentially be visible from an orbiting survey camera.

Invertebrate Fossil Record

More than 35 different phyla are alive today, and most of these have representatives in today's ocean waters. There was a time in Earth's past, however, about a billion years ago, when animals were rare or absent in marine environments. This began to change suddenly around 600 million years ago, when fossils of animals and animal-like organisms begin to appear in the fossil record.

THE CLEMENTE FORMATION

The record of fossil animals before 600 million years ago is like the early fossil record of bacteria and stromatolites, namely, highly contentious. Most reported animal fossils from before this time, be they putative body fossils or trace fossils, are either pseudofossils or inaccurately dated. At about 600 million years old

(the rocks are not yet precisely dated, so this date is plus or minus 20 million years), marine sedimentary rocks in northern Mexico provide evidence for the oldest diverse animal communities. This rock unit, called the Clemente Formation, is Late Proterozoic in age and has yielded both body fossils and trace fossils of what are either the earliest animals or their close relatives or both.

Some of these Mexican fossils belong to a group known as the Ediacarans. Ediacarans are a puzzling group of marine fossil

Above: A fossil heart urchin, member of the animal phylum Echinodermata, a group that became very successful beginning in the mid-Mesozoic. Right: A starfish, a modern representative of the animal phylum Echinodermata. Top left: This beautiful modern urchin, Astropyga magnifica, was collected on Diaphus Bank, in the Gulf of Mexico.

organisms that have bodies shaped like leaves or pancakes. Some of the Ediacarans reached large size, up to a meter or more in some cases. Although there are similarities between the various types of Ediacarans, it is not at all clear whether they represent variations on a single, very strange kind of extinct organism or rather represent anomalous representatives of more familiar animal phyla such as mollusks and arthropods.

Nonetheless, the Clemente Formation provides evidence of both Ediacarans and animal trace fossils appearing simultaneously, strongly suggesting that the earliest animals and the first Ediacarans were part of the earliest communities of complex multicellular life.

A CAMBRIAN EXPLOSION

Some tens of millions of years later, at the Precambrian-Cambrian boundary, a threshold is crossed that represents the primary division in the geological record. This boundary, dated to approximately 542 million years ago, is the moment in geologic time when fossils of familiar general types of animals become abundant for the first time. The first snails, clams, echinoderms, trilobites, and brachiopods appear as abundant fossils, and by the Ordovician and later times they are littering the shallow seafloor with their shells and, at least in the case of echinoderms, creating entire formations of bioclastic limestone primarily of their skeletal debris.

SNAILS AND ECHINOIDS

Generally speaking, invertebrate marine animals expand in diversity throughout the Paleozoic, but take a drastic drop in numbers at the Permo-Triassic mass extinction boundary that begins the Mesozoic. From the few survivors, diversity is restored to the oceans by the Middle Mesozoic. The end of the Mesozoic, although witnessing a severe extinction event that kills the dinosaurs, has a relatively smaller influence on invertebrate marine animals living on the sea-floor. For example, an advanced Mesozoic group of snails known as the neogastropods scarcely notices the Cretaceous-Tertiary mass extinction and continues its trajectory of diversification well into the Cenozoic. Similar statements could be made about the irregular echinoids, a group closely related to sea urchins, which becomes highly successful in the Cenozoic because of their ability to burrow and feed at the same time using myriad short, mobile spines covering their plated bodies.

A large collection of fossilized brachiopod shells. Along with echinoderms and other life-forms, the first abundant brachiopod fossils appear during the Cambrian Explosion around 542 million years ago.

Origins of Skeletons

Skeletons are formed by a biochemical process known as biomineralization. Most biotically formed minerals have geological importance, because mineralized hard parts of organisms have a high likelihood of being preserved in the rock record. There are, however, exceptions to this general rule. The organic material called sporopollenin, a constituent of pollen grains, and the resting cysts of marine microbes called dinoflagellates, can last for billions of years intact in sedimentary rocks that have not been overheated. And some skeletal minerals, such as celestite (a crystalline form

of strontium sulfate used by a group of marine microbes called acantharians), dissolve so quickly that they never have a chance to accumulate on the seafloor to become fossils.

MICROBIAL SKELETONS

Microbial fossils on Earth go back in time more than three billion years; however, diverse types of microbial skeletons do not appear until relatively late in Earth history. The oldest biomineralized structures that can be recognized in the fossil record are tiny magnetite crystals thought to have been secreted by magnetotactic (magnetic-

field sensitive) bacteria as an aid to navigation by means of Earth's magnetic field. These fossil crystals are thought to be distinctive for their unusual truncated hexoctahedral crystal form, but there is some controversy associated with their recognition. Some geologists argue that such crystals could have been formed by inorganic means. More complex microbes use iron oxides as well, such as protists including *Anisonema* that, like the magnetotactic bacteria, form chains of magnetite crystals, probably to help with orientation in a magnetic field. A variety of marine protists

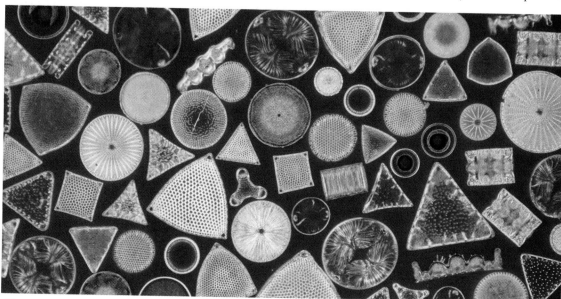

Above: Light micrograph of assorted species of fossil diatom shells. Top left: Clear gray-blue lustrous celestite crystals.

are able to use, in addition to celestite, the minerals calcite (calcium carbonate), aragonite (a needlelike form of calcium carbonate), barite (barium sulfate), gypsum (hydrated calcium sulfate), and, most interestingly, hydrated silica, or opal. Diatoms in particular are known for their opaline skeletons. Note the abundance of sulfate minerals (celestite, barite, gypsum) in the list of microbial biominerals. This is likely a reflection of the abundance of sulfate in seawater, second only to the sodium and calcium ions in solution.

The limestone islands of Palau show their distinct uplifted mushroom shape.

ANIMAL SKELETONS

Animals are not able to form quite as many biominerals as can the microbes (particularly the protists), but for what they lack in variety the animals make up for in bulk. Entire limestone reefs, for example, are formed primarily (but not exclusively) due to the biomineralizing activities of animals such as hermatypic (mound-forming) corals. Sponges are able to make their skeletons (formed of pointy spicules) out of either opal or calcium carbonate. Mollusks, likewise, can form skeletons out of several varieties of calcium carbonate (such as aragonite for the inner shell and calcite for the outer shell in mussels) and magnetite. The magnetite in mollusks forms the tiny teeth on the molluscan beltlike tongue called the radula. Mollusks such as grazing snails use their radula for scratching into rock surfaces to excavate and ingest algae and bacteria that thrive between the

Magnetite forms the tiny teeth of the grazing snails, such as this abalone.

outer pore spaces of the rock surfaces. Limestone islands in tropical regions have developed mushroom shapes as a result of gradual grinding away of the island at water level where the algal growth is most concentrated. In this case, one type of biomineralization (iron-bearing radular teeth, or sclerites) grind away at another type of biomineral (the calcite and aragonite of the limestone island). Animals can also make skeletons out of calcium phosphate (apatite and its derivatives such as hydroxyapatite). Apatite skeletons first become abundant during the Cambrian evolutionary event at 542 million years ago, when forms such as *Protohertzina* and *Lapworthella* appear. Phosphate mineralization is still a feature of the animal skeleton today, and can be found in our own bones and teeth.

SYSTEMA NATURÆ
IN QUO
NATURÆ REGNA TRIA,
SECUNDUM.

Convergent and Divergent Evolution

In order to study evolution, it is important to have an agreed-upon system of classification for both modern and ancient or extinct organisms. Two systems are currently in use, the Linnaean system and the cladogram system.

LINNAEAN CLASSIFICATION

In Linnaean classification, each type of organism that constitutes (or is thought to have constituted) a breeding population of organisms is assigned to a species. A species is therefore the basic taxonomic unit, or taxon. These species in turn are assigned to genera, and each such genus must contain at least one species. Each species is given a genus and species designation, such as our own, *Homo sapiens*. Together, genera and species are called in the Linnaean system the lower taxa. The Linnaean system has higher taxa as well. Genera are grouped into families, families are grouped into orders, orders into classes, classes into phyla (singular *phylum*), and phyla into kingdoms. This system has been in use for many years as a means of denoting evolutionary relationships, and it has served well for this purpose, but it does have some weaknesses. For example, at times, key evolutionary branch points tend to fall between taxonomic categories, and this has led to the proliferation of ad hoc taxa such as superfamilies, suborders, superphyla, and the like. Such proliferation can lead to a confusion of complicated terminology, all due to the fact that some of the key evolutionary relationships and branch points were not well understood at the time that the standard higher taxonomic categories were first defined.

CLADISTICS

The science of cladistic analysis was created as an alternative to Linnaean classification. In cladistic analysis, a series of key traits or characters are selected to define branching points in an evolutionary sequence. For example, the character "body hair" may be used as the key trait to distinguish mammals from reptiles. All mammals have

Above: Carl von Linné or Carolus Linnaeus (1707–78) is often called the Father of Taxonomy. Top left: Title page of Systema Naturae, *Linnaeus's classification of living things.*

NODE-BASED	A	B	Z
STEM-BASED	A	B	Z
APOMORPHY-BASED	A	B	Z

In the cladistic system of taxonomy, there are three ways to define a clade: node-based, the most recent common ancestor of A and B and all its descendants; stem-based, all descendants of the oldest common ancestor of A and B that is not also an ancestor of Z; and apomorphy-based, the most recent common ancestor of A and B possessing a certain derived character, and all its descendants.

hair, and no reptiles have hair. An alternate trait to use might have been, say, "suckle young." All mammals suckle young, whereas no reptiles suckle their young. With a collection of three or more different types of organisms, a cladogram can be constructed to render the evolutionary relationships based on this use of what are called derived characters.

CONVERGENT EVOLUTION

At times during the evolutionary process, different types of organisms can settle on the same solutions to various problems of life. These similar solutions can be manifested as similarities in body morphology or behavior. Such resemblances are said to be due to the process of convergent evolution. Convergent evolution can be a serious problem for cladistic analysis, because a false cladogram that is not representative of a true family tree (or phylogenetic relationships) results if convergent characters are used as key traits. For example, "warm bloodedness" (homeothermy) might be a poor choice as a distinguishing character in a cladogram that included both birds and mammals, because birds and mammals developed homeothermy independently in their separated evolutionary

lineages. Therefore, a goal of cladistic analysis (a goal it shares with Linnaean classification) is to describe the course of divergent evolution—in other words, to identify the key branching points in the evolutionary sequence. Convergent evolution has been shown to be so widespread in the evolutionary sequence that it seems that different lineages are constrained by their habitats to converge on the same solutions to a variety of biological challenges.

Birds, along with mammals, are warm-blooded. Because this trait was developed independently in both families, it is known as convergent evolution.

Vertebrate Fossils

Vertebrate animals do not constitute an entire phylum, but are the backboned subset of the animal phylum Chordata.

EARLIEST CHORDATES

The earliest members of phylum Chordata are known from Early and Middle Cambrian rocks of China and northwestern Canada. These fossils, such as *Pikaia* from the Middle Cambrian Burgess Shale, are boneless creatures with evidence for chevron-shaped muscle sets (as seen in a modern fish fillet) and a stiffening rod with a spiral wrap called the notochord. *Pikaia* and its relatives from China are entirely soft-bodied and require special conditions of preservation in order to appear as fossils. Chordates with hard parts

Above: The Burgess Shale fossil quarry, discovered near Field, British Columbia, in 1909 by American paleontologist Charles Doolittle Walcott. Top left: Amphibians, like this salamander, were the first land vertebrates.

appear later in the Cambrian, and a series of small plates and denticles belonging to early fishlike creatures appear in the fossil record. One of these that had dentine in its hard parts, as vertebrates do in their teeth, was the conodonts, a close relative of vertebrates. Conodonts were eel-shaped marine predators with sharply pointed grasping spines or teeth used to subdue smaller swimming animals. Conodonts are important index fossils from the Cambrian to the Triassic, for they occur in a wide variety of marine habitats (thanks to their swimming lifestyle that would

The Burgess Shale fossil Pikaia *is one of the earliest known representatives of the animal phylum Chordata. As such, it is a relative of the ancestor of all vertebrate animals, including of course humans.* Pikaia *may have been a swimming filter feeder.*

carry them over a variety of seafloor sediment types). They underwent very rapid evolution and, like certain types of trilobites in the Early Paleozoic, are characteristic for particular slices of geologic time in the Middle and Late Paleozoic.

AGE OF FISH

By the Silurian period, a variety of jawless (agnathan) and jawed (prognathan) fish were living in marine habitats, but the real explosion in diversity of fish does not come until the Devonian period. At that time, fish began to move into estuarine environments that were newly enriched in nutrients derived from the first bloom of vascular plants on the continents. These nutrients evidently drove rapid evolution in a variety of fish groups, and fish as diverse as lungfish, lobe-finned fishes, carp and trout ancestors, and other types of bony

fish appear in these nearshore aquatic habitats. Although nutrient and food rich, these habitats came with a price for fish, namely, bacteria were so active in these waters that they depleted dissolved oxygen content of the water to suffocation levels for fish that breathe with gills. Therefore, most of the nearshore Devonian fish were able to breathe with lungs as well as gills. Later, lung breathing was lost in many of the fish types as they moved back into deeper waters. For example, the swim bladder in a goldfish is an evolutionary derivative of the Devonian lung in the goldfish lineage. In coelacanths, a group of lobe-finned fishes thought to have gone extinct in the Mesozoic but found living off the coast of Africa in the 1930s, the lungs were evolutionarily transformed into internal fat storage centers, analogous to the blubber of a whale.

TETRAPODS

Another group of Devonian lobe-finned fishes evolved into the tetrapods, the group that gave rise to all of the four-footed vertebrates on land. The first of these were amphibians, animals that used modified fins to travel in shallow mud and, eventually, over dry land. At first tied to the water for reproduction via gelatinous, moisture-requiring eggs, one group of amphibians developed into the first reptiles, animals that lay hard-shelled eggs on dry land. This proved to be such a successful innovation that it is still used today by most reptiles, birds, and even a few mammals such as the duckbilled platypus and the spiny echidna of Australia.

Coelacanth swimming. Coelacanths are a surviving group of lobe-finned fishes, the fish that gave rise to tetrapods such as reptiles and mammals.

Vascular Plant Fossils

The fossil record of vascular plants is varied. There are distinctive chemical fossils, such as pristane and phytane, and entire formations of a rock called coal that is so unusual as to not really have appeared in any quantity in the rock record until the appearance of abundant vascular plants in the Late Devonian.

PLANTS AND NUTRIENTS

It is no coincidence that this is approximately the same time that fish underwent their critical (for us) adaptive radiation. Evidently, the fish were feeding on arthropods fed by the rapidly expanding plant communities. Paleontologist Richard K. Bambach has argued, in fact, that the expansion of the plant communities on land led to major nutrient increases in the ocean, as newly flourishing land biotas leaked nutrients out of the crustal rocks and into rivers, which of course made their way to the sea.

HYPERSEA

From rather obscure beginnings among the green algae, vascular plants underwent a dramatic radiation in the Middle Paleozoic, leading to forms such as sphenophytes (horsetails) and lycophytes (club mosses). Plants able to colonize dry upland habitats appear in the fossil record by the Late Paleozoic, and in the Late Mesozoic, the adaptive radiation of flowering plants (angiosperms) rivals in intensity that of the Paleozoic. Part of the reason for the success of all these vascular plant life-forms is their association with soil fungi in a specialized type of tissue known as the mycorrhiza. This vascular plant-fungal symbiosis has led to a tremendously important nutrient exchange that has been in place ever since the Silurian and the first relatively large plants such as the Australian lycophyte *Baragwanathia*. In plants with this kind of symbiosis (most modern vascular plants have this, with a few exceptions such as Chinese cabbage), the fungus delivers to the plant abundant mineral nutrients absorbed from soil minerals, and the plant delivers to the

Left: Club moss found growing near the thermal vents at Craters of the Moon in Taupo, New Zealand. This modern lycophyte plant had much larger relatives in the Carboniferous period. It does not seem to be bothered by the heat, steam, and sulfur of its geothermally influenced habitat.
Top left: Pecopterix *fossil leaf impression from the Carboniferous.*

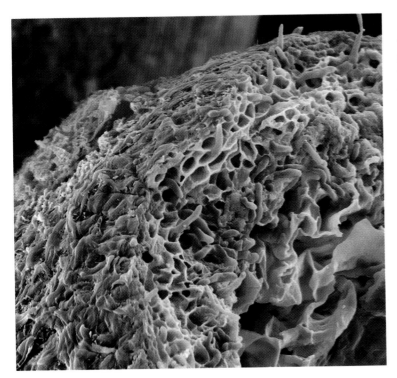

Glomus. After the time of Rhynie, vascular plants and their fungal consorts began to enlarge and to attain the shapes and sizes of modern trees. The familiar tree form does not represent a single plant lineage but rather has evolved multiple times since the Devonian. Whenever vascular plants begin to form stacked arrays of vascular (fluid-conducting) cells called tracheids, wood forms, and with wood-forming ability plants can fairly rapidly grow into trees that can develop a forest canopy.

Left: The mycorrhiza seen in this colorized scanning electron micrograph represents a symbiotic association between a soil fungus and vascular plant roots. This symbiotic connection is essential for the plant, for the fungus is able to absorb mineral nutrients otherwise unavailable to the plant. Below: A thin section through Rhynie chert shows fossilized stems of the plant Rhynia major. Study of Rhynie chert has uncovered some of the earliest known evidence for a symbiotic relationship between plants and fungi.

fungus its excess sugars produced by photosynthesis. This symbiotic system has been called Hypersea, and the process of nutrient flow upward from the crust has been called hypermarine upwelling.

THE RHYNIE CHERT

The first direct evidence for this fungal-vascular plant connection is from the 400-million-year-old Rhynie chert, a terrestrial chert from near Aberdeen, Scotland. Chert is a chemical sedimentary rock formed by tiny quartz crystals that precipitate from an aquatic gel on the seafloor, or in the case of the Rhynie chert, within a marshy wetlands environment. Freshwater cherts such as that at Rhynie are relatively rare, but the Rhynie chert more than makes up for this by providing spectacular fossils of early vascular plants such as *Rhynia*; fungi and their reproductive spores preserved within the fossil vascular plants themselves; and fossils of tiny animals that fed on the plants and fungi. This spectacular deposit (now, ironically, covered by a cow pasture) provides convincing fossil evidence that the vascular plant–fungal symbiosis was well established by 400 million years ago. The fossil fungi are so well-preserved that they can be assigned to modern soil fungi genera such as *Sclerocystis* and

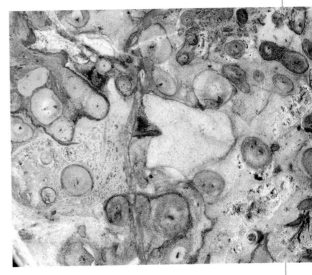

History of Diversity

It seems logical to infer that life on Earth developed from one species or a few and that the diversity of life forms has increased ever since. Nonetheless, the actual pattern of this diversity increase is not known with certainty, although it has been a topic of intense interest to geologists and paleontologists almost since the beginnings of paleontology as a modern science.

PHILLIPS'S PLOT

In his 1860 book, *Life on Earth*, geologist John Phillips made a preliminary estimate of the diversity of life-forms on Earth through geologic time. His plot showed diversity being low in the Paleozoic, and then dropping to very low levels at the Permo-Triassic. Mesozoic diversity climbed to a high level,

more than twice that of the Paleozoic, and then dropped even more precipitously at the Cretaceous-Tertiary boundary. Diversity climbs again in the Cenozoic to reach a level that is at least half again as high as the Mesozoic maximum.

THE SEPKOSKI CURVE

Phillips's diversity plot was not much remarked upon for more than a hundred years. But in the 1970s, the attentions of geologists once more returned to the question of the diversity of life through time. University of California evolutionary biologist James Valentine published a plot at this time that had much in common with Phillips's plot, at least at the lower taxonomic levels such as families and genera. Using a vast compendium of paleontological data, first at the family level and later at the genus level, University of Chicago paleontologist Jack Sepkoski (1948–99) attempted to quantify the diversity curve. He felt that he could represent three separate faunas that expanded and contracted over the last half billion years. Each of the three faunas began in the early Cambrian, and Sepkoski posited that each had its own internal dynamic, its

own rate of exponential expansion, and, most important, its own equilibrium or maximum level of diversity. For any given fauna, this diversity level would remain constant unless perturbed by a mass extinction or by extinctions caused by competition with the members

Left: A lower Paleozoic trilobite, a marine creature that appeared 600 million years ago and flourished for 350 million years. Above: John Phillips. Top left: Modern cephalopod (Sepia, the cuttlefish), descendant of cephalopods that first appeared during the Cambrian.

of subsequent faunas still in their expanding, exponential phase. Sepkoski concluded, in agreement with Valentine and Phillips, that diversity had indeed increased over time and that the increase was especially marked in the Cenozoic. An interesting by-product of Sepkoski's research on the evolutionary faunas was that the background rate of extinction showed a statistical decrease from slightly more than six families going extinct per million years in the Cambrian to slightly less than six families going extinct per million years by the late Cenozoic. Also of interest were suggestions in the data of a 26-million-year periodicity in extinction events; the significance of this possible pattern is still being debated today.

Above: An extinction event occurs when there is a sharp decrease in the number of species in a relatively short period. One such event, possibly caused by a massive meteor impact, wiped out the dinosaurs. Left: Graph showing the "Sepkoski curve," a plot of marine diversity at the genus level through geologic time.

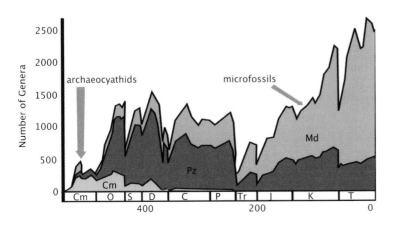

Dinosaurs: A Case Study

The dinosaurs provide a wonderful case study for tracking the migration and distribution of an important group and for understanding the evolution of a major taxon, or clade, through geologic time. Questions involving the origin and distribution of dinosaurs, the reason or reasons for their demise, and their relationship to modern birds are among the most fascinating in modern paleontology.

DINOSAUR ORIGINS

Dinosaurs are currently thought to range from the Late Triassic to the Late Cretaceous, a span of approximately 163 million years. The earliest dinosaurs clearly have a Gondwanan origin, and include species such as *Eoraptor lunensis* and *Herrerasaurus ischigualasto* from

Above: This fossilized theropod dinosaur footprint, dated to the Lower Jurassic period, was found in St. George, Utah. Right: An artist's depiction of a Stegasaur *and a* Brachiosaur *of the Late Jurassic. Top left: Bones from Drumheller, Canada's "Dinosaur Valley."*

the Ischigualasto Formation of Brazil and *Staurikosaurus pricei* from the Santa Maria Formation of Brazil. Late Triassic rocks of Argentina have also produced *Marasuchus*, an archosaurian form with long hind legs that shares many characteristics with true dinosaurs. Dinosaurs are descended from a group of Early Mesozoic reptiles known as the archosaurs, and a key adaptation that seems to set dinosaurs apart from most of their archosaur relatives is elongated hind limbs that allow a bipedal stance. Bipedal locomotion and its link to rapid motion on land may be the key to subsequent dinosaur diversification.

DINOSAUR DIVERSIFICATION

From the Triassic to the Cretaceous, three centers of dinosaurian diversification can be identified on a Mesozoic world map. During dinosaur ascendancy, the world went from a supercontinental configuration (Pangea) to a world in the Cretaceous with north-south oceans and isolated continents that very much resembles the modern world. The first center of dinosaur diversification is evidently in Gondwana (particularly what is now South America, southern Africa, and

Dinosaurs:
A Case Study (continued)

Antarctica). Prosauropods such as *Riojasaurus* and *Massospondylus* appear in Late Triassic rocks of Argentina and South Africa, respectively, and the oldest prosauropod is *Saturnalia tupiniquim* from Brazil. Ornithopod dinosaurs such as *Abrictosaurus* and *Heterodontosaurus*

appear in the Early Jurassic of South Africa. The large carnivorous dinosaur group Tetanurae, a group that includes *Tyrannosaurus rex*, has as its earliest known representative *Crylophosaurus* from Antarctica.

The second main center of dinosaur diversification is in Europe. Predatory theropods (ceratosaurs; Late Triassic), quadripedal herbivores (ankylosaurs; Middle Jurassic), and bipedal herbivores (pachy-

cephalosaurs; Early Cretaceous) make their first appearances in Europe. The third center of dinosaur diversification is China, where the first true sauropods appear in the Late Triassic, the first stegosaurs in the Middle Jurassic, and finally the first ceratopsians appear in the Early Cretaceous.

BIRDS OF A FEATHER
Ever since dinosaur tracks were first discovered in 1802 by

Left: Archaeopteryx, *the earliest-known fossil bird. Above: Fossilized skull of* Tyrannosaurus rex. *Opposite right: Artist's conception of a theropod—the direct ancestors of modern-day birds. Top left: Prehistoric flying reptiles, such as the pteranodons, were not the ancestors of modern-day birds.*

Pliny Moody in South Hadley, Massachusetts, and described as the tracks of "Noah's Raven," connections have been proposed between dinosaurs and modern birds. The now obsolete view of dinosaurs as overgrown, sluggish reptiles has given way to a view of dinosaurs as highly active and successful animals with birdlike behavioral characteristics such as nesting and care of young. Discovery of an Early Cretaceous small Chinese theropod, *Microraptor,* with four feathered limbs has decisively added to a keen appreciation of the dinosaurian traits (such as teeth and forelimb claws) as seen in the earliest known fossil bird, *Archaeopteryx.* Dinosaurs are traditionally classified into two main groups, the "lizard-hipped" Saurischians and the "bird-hipped" Ornithschians. In a somewhat ironic case of convergent evolution, birds are direct descendants of theropod saurischians rather than being descended from ornithschians. Most dinosaur paleontologists today would claim that dinosaurs are not extinct; they live among us as the theropods we know as birds.

Notable Names in Geology

Thales
(c. 640–546 bce). Greek philosopher. He was the first to make the scientific inference that life originated in water and subsequently moved onto land.

Aristotle
(384–322 bce). Greek philosopher and naturalist. Aristotle made the first concerted effort to classify living organisms according to their characteristics or morphological traits.

Albertus Magnus

Albertus Magnus
(1193/1205–80). Saint and doctor of the Roman Catholic Church. Albert the Great wrote an important treatise, the *Book of Minerals*, that dealt with the challenging subject of ore mineral genesis. For example, he was able to distinguish placer gold deposits from vein ore.

Leonardo da Vinci
(1452–1519). Italian artist and inventor. Da Vinci demonstrated that fossil shells in mountain areas were once marine, and in his Codex Leicester hypothesized a mechanism by which sea level remains constant.

Nicolaus Steno (Nils Stensen)
(1638–86). Danish naturalist and clergyman. Steno founded modern stratigraphic principles based on his studies of the geology of Tuscany, Italy. Steno also convincingly demonstrated the similarity between fossil and modern shark teeth, indicating that the former had once been living.

Mikhail Vasilevich Lomonosov
(1711–65). Russian naturalist. Lomonosov was among the first to use actualistic principles in geology. Actualism is the concept that the "present is the key to the past." In other words, many past processes operated in the same way as they do today.

Jean-Étienne Guettard
(1715–86). French geologist. Guettard demonstrated the igneous nature of the ancient volcanoes in the Auvergne district, France, and published the first geologic map in 1746.

James Hutton
(1726–97). Scottish geologist. Basing his work on the famous Siccar Point unconformity (stratal gap), Hutton became the first person to grasp the immensity of geologic time ("deep time"). Hutton also described the organismlike characteristics of the Earth system, what is now called geophysiology.

Nicolaus Steno

Abraham Gottlob Werner
(1749–1817). Prussian geologist. Charismatic lecturer who in 1777 published a book that outlined a globally applicable stratigraphic scheme that encompassed all geologic time.

Jean-Baptiste Lamarck
(1744–1829). French naturalist. He inferred that certain biological traits can be acquired rather than inherited, a view that has been validated by symbiosis theory. His 1801 book, *Hydrogeology*, pointed out the signal importance of life for geological processes.

Georges Cuvier
(1769–1832). French naturalist. Cuvier used comparative anatomy to reconstruct ancient animals, to show that certain creatures had gone extinct, and to identify, in a dramatic demonstration, marsupial remains in the fossil record.

Georges Cuvier

William Smith

(1769–1839). British geologist. Smith published the first national geologic map that showed comprehension of the sequence of strata. He used fossils and rock type similarities to demonstrate the value of these sequences for correlating strata.

Charles Lyell

(1797–1875). Scottish geologist and barrister. Lyell established the uniformitarian (as opposed to catastrophist) principles of geology with the publication of his 1830 book, *Principles of Geology.*

William Smith

Jacques Joseph Ebelmen

(1814–52). French mining engineer and chemist. Ebelmen introduced the modern chemical formula in use today and demonstrated the importance of plant activity for enhancing the weathering rates of carbonate and silicate rocks and soils.

Corneille Jean Koene

(c. 1817–c. 1865). Belgian chemist. Koene was first to infer that the carbon dioxide content of the atmosphere had undergone a dramatic decrease over geologic time as a result of the utilization and burial of carbon by plants and their decomposing remains.

Eduard Suess

(1831–1914). Austrian geologist. Suess identified the supercontinent Gondwana in 1861 and the Mesozoic Tethys Ocean in 1893. In his book *The Face of the Earth,* Suess introduced the concept and term "biosphere."

Adam Sedgwick

(1785–1873). British geologist. Sedgwick defined the Cambrian period in 1855 and emphasized that an accord must exist between the truths of science and the truths of faith—"truth must at all times be consistent with itself"—a theme also emphasized by American geologist and first president of Amherst College, Edward Hitchcock (1793–1864).

Roderick Murchison

(1792–1871). British geologist. Murchison defined the Silurian period, and made an economically successful correlation between Paleozoic strata (of differing rock type) between Britain and Russia.

Louis Agassiz

(1807–73). Swiss-born American paleontologist and geologist. Agassiz conducted pioneering research on fossil fish and convinced his contemporaries that a major ice age had (in geological terms) recently influenced the northern continents.

Charles Lyell

Louis Agassiz

Charles Darwin

(1809–82). British naturalist. Darwin, justly famed for expanding the evolutionary ideas of his grandfather Erasmus Darwin into the theory of evolution by natural selection, was a superb geologist who contributed to topics as diverse as the origin of atolls and the soil-churning abilities of earthworms.

Notable Names in Geology (continued)

Edward Drinker Cope

Edward Drinker Cope

(1840–97). American paleontologist. A participant in what has come to be known as the Bone Wars, Cope searched Cretaceous strata in western North America and discovered 56 species of dinosaurs. Cope's law is the concept that evolutionary lineages increase in body size over geologic time.

Thomas Chrowder Chamberlin

(1843–1928). American geologist. Chamberlin, with astronomer Forest R. Moulton, defined the planetesimal hypothesis for the formation of Earth. Chamberlin also wrote forcefully about the process of scientific thought in his famous essay "Method of Multiple Working Hypotheses."

Charles Doolittle Walcott

(1850–1927). American paleontologist. Secretary of the Smithsonian Institution and the United States Geological Survey, Walcott is famed for his discovery of the soft-bodied Cambrian fossils of the Burgess Shale from near Field, British Columbia, Canada.

Andrija Mohorovicic

(1857–1936). Croatian geophysicist. After a 1909 earthquake near Zagreb, Mohorovicic discovered primary and secondary seismic waves. He went on to delineate the discontinuity that bears his name (the Moho Discontinuity) between the Earth's mantle and crust.

Louis Dollo

(1857–1931). Belgian paleontologist. Dollo gained fame for his excavation of the deposit of complete *Iguanodon* skeletons in Bernissart, Belgium, and for Dollo's law, the concept that evolutionary change is irreversible.

Vladimir Vernadsky

(1863–1945). Russian geochemist. Vernadsky developed the modern concept of the biosphere as a geocatalytic entity that is ultimately responsible for most of the surface geology of Earth. Vernadsky calculated the "speed of life."

Charles Doolittle Walcott

Louis Dollo

Joseph Barrell

(1869–1919). American geologist. Barrell used a combination of stratigraphic analysis and radiometric dating to establish the link between relative and absolute dating techniques, respectively, used today in geology.

Mignon Talbot

(1869–1950). American paleontologist. Talbot published the first largely accurate dinosaur skeletal reconstruction in her 1911 description of *Podokesaurus holyokensis* in the *American Journal of Science*.

Alex du Toit

(1878–1948). South African geologist. In his book *Our Wandering Continents*, du Toit used his vast knowledge of the southern continents to argue persuasively that Gondwana had broken apart by continental drift.

Milutin Milankovitch

(1879–1958). Serbian astrophysicist. Milankovitch is best known for the theory that bears his name, relating changes in climate to cyclic variations in Earth's orbital and rotational parameters.

Alfred L. Wegener

Alfred Lothar Wegener

(1880–1930). German meteorologist and geologist. Wegener strongly advocated the theory of continental drift with the publication of his book *The Origin of Continents and Oceans*. Wegener, who named the supercontinent Pangea, died in Greenland during an expedition whose goal was to advance drift theory.

Pierre Teilhard de Chardin

(1881–1955). French Jesuit priest and geologist. Teilhard was a geological explorer and paleontologist famed for his work in China. He emphasized evolutionary convergence in his influential, posthumously published book *The Human Phenomenon*.

Roy Chapman Andrews

(1884–1960). American adventurer. A likely prototype for Indiana Jones, Andrews led an expedition to the Gobi Desert in search of human origins and ended up finding the first-known dinosaur eggs, now categorized as belonging to the theropod *Oviraptor*.

Raymond Dart

(1893–1988). Australian paleoanthropologist. In 1924, Dart discovered a juvenile skull of *Australopithecus africanus* (the Taung skull) in a box of fossiliferous limestone shipped to him from a quarry in Taung, northwestern South Africa.

Harry Hammond Hess

(1906–69). American geologist. Known for his classic work on peridotite complexes and his studies of the mineral pyroxene in lunar samples, in 1960 Hess hypothesized correctly that mantle convection cells were the driving force for continental drift and seafloor spreading.

Luis Walter Alvarez

(1911–88). Spanish-born physicist and Nobel laureate. Alvarez discovered the iridium anomaly at the Cretaceous-Paleocene

Roy Chapman Andrews and companion

Pierre Teilhard de Chardin

boundary, and in 1980 proposed the asteroid impact theory for the end-Cretaceous mass extinction with his geologist son Walter Alvarez.

Preston Ercelle Cloud Jr.

(1912–91). American geologist and Precambrian paleontologist. After studying the extensive banded iron formations of the Precambrian, Cloud hypothesized that they represented iron oxides formed in marine water, sent to the seafloor as rust, and formed by oxygen released by photosynthesis.

Marie Tharp

(1920–2006). American marine geologist. Tharp discovered the V-shaped notches along the crests of mid-ocean ridges, and her global maps of the geology of the seafloor established plate tectonic theory.

Milestones in Geology

Thales

1150 BCE
The Turin Papyrus survives as the earliest geological map.

600 BCE
Philosopher Thales infers that life originated in water.

332 BCE
Philosopher Aristotle classifies living organisms.

250 BCE
Archimedes discovers isostacy.

1245
Albertus Magnus distinguishes placer gold from vein gold.

1667
Steno establishes the principles of stratigraphy.

Fossiliferous stone from Hooke's Micrographia

1703
Robert Hooke suggests that fossils might be useful for dating, just like old Roman coins, implying that some lifeforms had gone extinct.

1746
Jean-Étienne Guettard makes the first geological map since the Turin Papyrus.

1777
Abraham Gottlob Werner establishes a global stratigraphy.

1785
James Hutton discovers "deep time."

1801
Strata Smith uses faunal succession to make a geological cross-section of Britain.

1802
Pliny Moody finds dinosaur trackways ("Noah's Raven") in South Hadley, Massachusetts.

Iguanodon

1821
British amateur Mary Mantell finds the first dinosaur bones, named *Iguanodon* by her husband Gideon Mantell.

1830
Charles Lyell publishes *Principles of Geology.*

1837
Louis Agassiz promotes the concept of a great ice age.

1838
Thomas Mitchell publishes the first geological map of Australia.

1845
Jacques Joseph Ebelmen invents modern chemical notation while studying mineral breakdown by weathering.

1855
Adam Sedgwick defines the Cambrian.

1856
Corneille Jean Koene discovers the major drawdown of atmospheric carbon dioxide by plants over geologic time.

1859
Charles Darwin publishes *On the Origin of Species.*

1861
Eduard Suess names Gondwana.

1878
First modern bathymetric map published, showing the Caribbean and the Gulf of Mexico, taken from soundings by the Coast Survey Steamer *Blake.*

The first modern bathymetric map

1890
Grove Karl Gilbert discovers ancient Lake Bonneville.

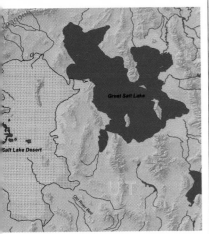

Glacial Lake Bonneville (in orange)

1895
Antoine-Henri Becquerel accidentally discovers radioactivity using uranium ore samples.

1900
Norman L. Bowen develops his Bowen's Reaction Series for understanding sequential crystallization in magmas.

1905
Thomas Chrowder Chamberlin and Forest R. Moulton propose the planetesimal hypothesis for Earth formation.

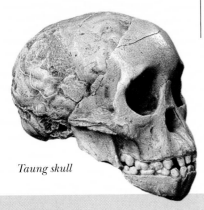

Taung skull

1909
Charles D. Walcott discovers the fossils of the Burgess Shale.

1909
Mohorovicic Discontinuity (Moho) discovered.

1911
Mignon Talbot accurately reconstructs the dinosaur *Podokesaurus holyokensis.*

1915
Alfred Wegener publishes *The Origin of Continents and Oceans;* Pangea named in a later edition of the book.

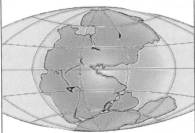

Pangea

1916
Ernst J. Öpik publishes his paper on the meteor theory for the origin of lunar craters.

1921
Aleksandr Oparin and J. B. S. Haldane agree that life must have evolved in a non-oxidized (reducing) atmosphere.

1923
Roy Chapman Andrews discovers dinosaur eggs in the Gobi Desert.

1924
Raymond Dart discovers the Taung skull in South Africa.

1925–27
The German ship *Meteor* surveys the Atlantic seafloor.

1935
Charles Palache publishes his classic paper on mineral paragenesis, describing the sequence of mineral formation in terms of crystal chemistry.

1947
Willard F. Libby discovers the radiocarbon dating technique, for which he will later win the Nobel Prize.

Seismic probability map

1948
Ulrich publishes the first seismic probability map.

1953
Samuel Epstein and his co-authors use oxygen isotopes in fossil foraminifera to track climate change.

1953
Stanley L. Miller, under the supervision of Harold C. Urey at the University of Chicago, conducts an experiment in which he simulates hypothetical conditions present on the early Earth.

Miller-Urey experiment

Milestones in Geology (continued)

Mauna Loa Observatory

1955

C. Patterson, G. Tilton, and M. Inghram calculate the age of the Earth.

1955

Bruce Hamon and Neil Brown develop the Conductivity-Temperature-Depth (CTD) meter, one of the most basic tools of modern oceanography. CTD meters allow scientists to measure how salinity (as indicated by conductivity) and temperature change with depth.

1956

German paleontologist Adolf Seilacher documents a profound change in trace fossil behavior across the Proterozoic-Cambrian boundary.

1956

British geophysicist Stanley Keith Runcorn uses paleomagnetism to determine ancient continental positions.

1957

James E. Lovelock invents the electron capture device for use in gas chromatography; this discovery revolutionizes atmospheric gas measurement.

1958

Precise atmospheric carbon dioxide measurements are begun at Mauna Loa Observatory, Hawaii, and prove essential for greenhouse climate research.

1960

Harry Hess links continental drift to mantle convection cells.

1961

B. W. Logan makes the critical connection between microbial mats and both ancient and modern stromatolites (Shark Bay, Australia).

1963

The new volcanic island Surtsey emerges off the coast of Iceland from eruptions on the mid-Atlantic ridge.

1964

Brian Harland identifies severe global glaciation in the Late Precambrian.

1965

E. S. Barghoorn and S. A. Tyler describe microbe fossils from the Precambrian Gunflint chert.

1968

Preston Cloud uses banded iron formations to date the oxygenation of the atmosphere.

1969

Apollo 11 space mission brings back lunar rocks.

Neil Armstrong collects lunar samples

1971

Mariner 9 becomes the first space probe to orbit another planet and discovers possibly water-cut channels on the Martian surface.

1973

Plate Tectonics and Geomagnetic Reversals sees print (Allan Cox, editor).

The island of Surtsey

Alvin

1977

After being attacked by first a swordfish (1967) and then a blue marlin (1971), the deep-sea submersible *Alvin* finally discovers the astonishing vent biotas at the Galapagos mid-ocean rift.

1977

Simon Conway Morris identifies priapulid worm fossils in the Middle Cambrian Burgess Shale.

1978

Walter Pittman III postulates a link between rates of seafloor spreading and global (eustatic) sea level.

1980

First inventory of global gas hydrate occurrences published as "United Stated Geological Survey Circular 825."

1980

Luis and Walter Alvarez unveil their impact theory for dinosaur extinction.

1983

A complete conodont fossil discovered in a museum drawer in Scotland.

1984

Allan Hills meteorite 84001, a sample of the ancient Martian surface, collected in Antarctica by snowmobile on the glacial ice surface.

1987

A complete specimen of the worm *Halkieria evangelista* described from Sirius Passet, Greenland, unlocks the secret to Cambrian small shelly fossils.

1990

Supercontinent Rodinia named.

1990

World Ocean Circulation Experiment begins.

1992

Joseph Kirschvink proposes Snowball Earth theory to explain severe Proterozoic glaciations.

1992

Paul C. Sereno describes the early dinosaur *Herrerasaurus ischigualasto* from South America.

1993

Lynn Margulis publishes *Symbiosis in Cell Evolution.*

1994

Hypersea: Life on Land published to explain life's transition to land and to extend Thales' line of inquiry.

AH84001

1995

Oldest animal fossils discovered in Sonora, Mexico.

2000

Space Shuttle *Endeavor* maps vast tracks of Earth with images, topological data.

2002

World Ocean Circulation Experiment ends.

2004

Cassini-Huygens space probe discovers dendritic channels on frozen surface of Titan.

Titan

Herrerasaurus ischigualasto

The Map That Changed the World

In 1801, an unknown British geologist named William Smith completed a map he labeled a "General Map of Strata in England and Wales." This map shows a series of parallel lines, representing lines of contact between different (but juxtaposed) layers of strata. The lines run from Devonshire in the southwest to York and the coast of the North Sea (Smith calls it the "German Ocean") on the northeast, and generally traces the main Jurassic outcrops of this part of Britain. This relatively simple map, showing a diagonal stripe running across the land, is a descendant of Jean-Étienne Guettard's 1745 map of the Cretaceous (chalk-bearing) strata of England and France. Smith's map thus truly became "the map that changed the world." Guettard's seminal map is entitled "The Mineralogical Map, Where One may Observe the Nature and the Disposition of the Terranes that Traverse France and England," and it shows the occurrences of coal, ores, and other types.

even shows the location of stone quarries and gold mining areas.

British Strata

Smith's "General Map of Strata" was not his first geological map creation; in 1799 he had published a map with a circular border of the region around Bath, England. By 1815, Smith had published his definitive map that delineates the strata of England, Wales, and Scotland. This multicolored production is directly comparable to the modern geologic map of the same regions, testimony to the accuracy and thoroughness of Smith's work. The map also shows a sea monster–like form rising out of

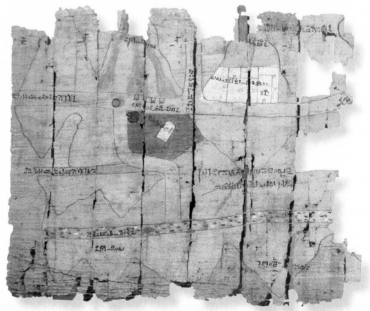

The Turin Papyrus

The maps of Guettard and Smith share an ancient antecedent. In 1150 bce, "the mapmaker" (a well-known ancient scribe) created what is the only topographic map to survive from ancient Egypt. Discovered by the French consul general between 1814 to 1821, the map portrays black sandstones, granites, serpentinites, pink volcanic rocks, and

Above left: William "Strata" Smith, often called the father of British geology.
Left: Guettard's mineralogical map. Above: Fragment of the Turin Papyrus.

the "German Ocean" (*Mer du Nord* for Guettard)—an early multicolored cross-section that shows the southeastward dip of the British strata.

Smith first became aware of the value of the ability to correlate strata from one area to another while he was making a valuation survey of coal mines on an estate in Somerset. At one of these mines, the Mearns Pit, Smith began to notice a regular succession in the fossils and the rock types that this stratigraphic succession contained. Smith's earliest known (1792) unpublished paper, an "Original Sketch and Observations of my First Subterranean Survey of Mearns Colliery in the Parish of High Littleton," contains these observations: "The stratification of the stones struck me . . . and [I] soon learned enough of the order of strata to describe on a plan the manner of working the coal in the lands I was then surveying." This work led directly to Smith's interest in using the orderly sequence of strata to map the rock bodies.

Slow Recognition

William Smith, a man of humble origins, was forced to wait many years before his work was acclaimed for the scientific advance that it really represented. Recognition did finally arrive, and in 1831 Smith was awarded the first Wollaston Medal of the Geological Society. Called William "Strata" Smith or simply Strata Smith by many, he eventually became known as the "father of British geology."

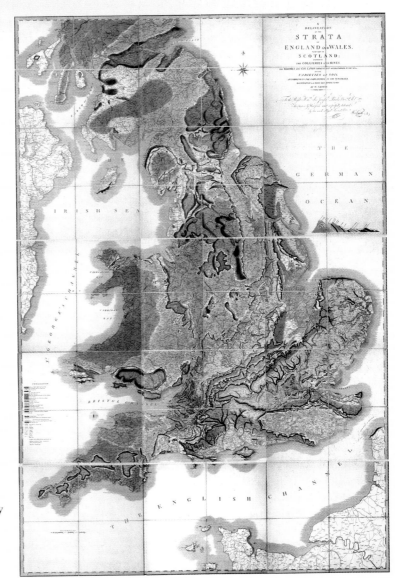

William Smith's ground-breaking "General Map of Strata in England and Wales."

The Mapping Program

From its beginnings in 1745, the geological map concept spread rapidly through Europe to North America, and to the rest of the world. Originally drawn to show the contacts between strata and the location of mines, geological maps accumulated additional symbols to denote fault traces, fold axes, attitude of strata (strike and dip symbols) and other significant features. High-technology aids such as digital altimeters and GPS have greatly assisted the process of making accurate geological maps in the field, although the map-making process still requires the guidance of a skilled field geologist. Strata Smith's map that changed the world is rapidly becoming the map that covered the world.

Geological Map of the World

The geological map of the world is the product of several centuries of zealous research by geologists throughout the world. In many respects it is an unfinished work, because many areas of the Earth—particularly parts of central Asia, Africa, and ice-covered Antarctica—remained to be mapped in sufficient detail to match the best geologic maps in other parts of the world.

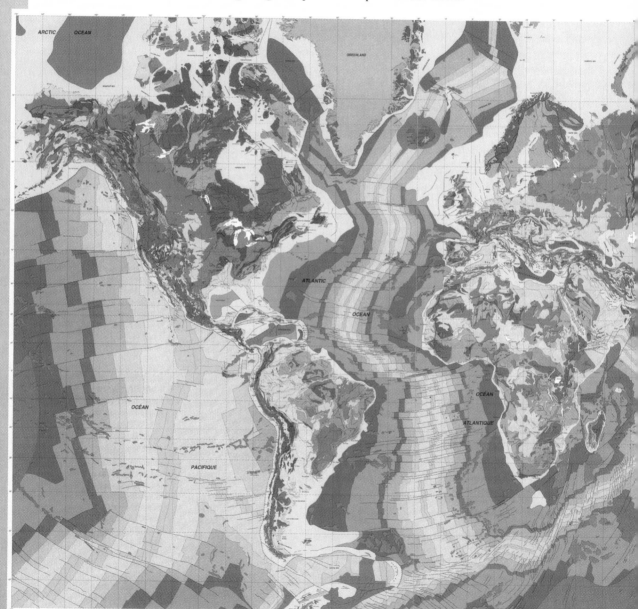

Map by the Commission for the Geological Map of the World

Ocean Basin

Perhaps the most striking feature of the geological map of the world is the pattern visible in the ocean basins of the world. The dramatic stripe patterns, symmetrical across the mid-ocean ridges, represent bands of mafic igneous rock

(primarily basalt) erupted in successive phases and at different points in geologic time dating back to the Jurassic. These stripes were first mapped by American marine geologist Marie Tharp. Most oceanic crustal rocks older than Jurassic have been destroyed by subduction, but in a few instances relatively small samples of older ocean crust have been preserved on the continents.

Continents

The geological maps of the continents are characterized by division of each continental map into an inner craton zone and an outer mobile belt. The craton consists of very ancient continental plates sutured together a billion or more years ago; Paul Hoffman has referred to the geologic map of the North American craton as the "United Plates of America." Cratons were once tectonically active areas that developed features such as the now highly deformed greenstone belts, but they have been quiescent for at least a billion years or even more. The occasional rift valley split is seen in the continental interior, but most of the tectonic activity occurs along the edges of the continents. Rare (but sometimes massive) earthquakes can be associated with the rift basins in continental interiors, but the majority of earthquakes are associated with mobile belts. These mobile belts are the venue for most of the mountain building that takes place on Earth, and wherever there is convergent tectonics (either subduction or continent-to-continent collision) in the vicinity of a continental margin, local tectonism expressed as orogeny (mountain building) is nearly certain to take place. When an island arc or microcontinent collides with a continental margin, as has occurred repeatedly along the southwestern coast of Alaska, new crust is added to the edge of the continent.

Time-Rock Units

Geologic time is divided up into eras, periods, epochs, and ages. With a geologic map in hand, we can refer to actual bodies of rock that were deposited during any particular period. The physical rock bodies therefore take on a time aspect, and we can talk about these rock bodies in the same way that we can talk about, say, all the red wine bottled in 2005. Hence, all the rocks deposited during the Mesozoic era belong to a body of rock known as the Mesozoic Erathem. An erathem is a time-rock unit. The rocks deposited during the Devonian period belong to a body of rock called the Devonian System. And all rocks deposited during the Miocene Epoch belong to the body of rock known as the Miocene Series. Note that any rock formed during the geologic time period in question, be it sedimentary or igneous, is included in the corresponding time-rock unit.

The Geological Time Scale

The geological time scale, representing as it does our scientific understanding of the age of the Earth, is one of the most important products of human effort and inquiry, and yet the time scale remains unfinished and will be the object of research for years to come.

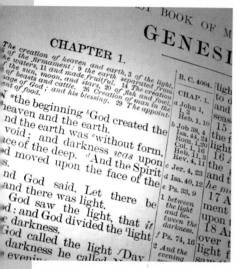

Biblical Origins

Before the development of geology as a scientific discipline, estimates of the Earth's age were based on biblical references. Although the Bible never explicitly states the age of the Earth, biblical texts were used to give a rough and relative chronology for the planet. Thus there was a biblical time of primordial chaos, a time before animals, and a time before humans. Early naturalists would divide history into halves based on the story of the universal flood.

Wernerian Chronology

The modern geological time scale can be said to have had an origin with the geological speculations of Abraham Gottlob Werner. Werner was a "neptunist" who believed that all or virtually all the rocks on Earth, including igneous and metamorphic rocks, were derived from precipitation from a universal ocean. Periodic inundations of the universal ocean represented for Werner successive iterations of Noah's Flood. This largely incorrect model nevertheless did allow Werner to create a global geological chronology, the first to be applied universally by scientists. Werner and his followers divided all geologic time into primary, transition, secondary, tertiary, and quaternary divisions.

Geological Periods

In the 1800s, more specific names began to be placed on the divisions in the geological time scale. Three geological eras came to be recognized: the Paleozoic, the Mesozoic, and the Cenozoic—rough equivalents to Werner's Transition, Secondary, and Tertiary. Geological periods, now known as subdivisions of the eras, came into use at this time as well. Some of these periods bear names reflecting places where the rocks are particularly well exposed, such as Devonian for Devonshire, England, and Permian for the Perm Basin in Russia. Others bear names of ancient peoples, such as Cambrian and Ordovician for Welsh tribes of antiquity. Still others refer broadly to the character of the rocks deposited at the time, such as Carboniferous for its abundant coal measures. Periods are further subdivided into epochs (such as Late Cretaceous or Eocene), and these are subject to further formal subdivision into ages.

Absolute Ages

For the early decades of use, the geological time-scale subdivisions were essential for determining the relative ages of strata. This relative-age dating was and still is put into practice primarily by examining the fossils in any particular sedimentary rock sequence. Absolute ages of ancient rocks, that is, their age in a particular number of years, remained out of reach (excepting some very broad estimates) until the advent of radiometric dating techniques, at which time it became possible to assign ages in millions of years to, say, the boundaries between geological periods. These values tend to shift slightly from time to time as dating precision advances in this high-tech part of geology.

Eon	Era	Period		Epoch	m.y.
Phanerozioc	Cenozoic	Quaternary		Holocene	
				Pleistocene	
					1.5
		Neogene		Pliocene	
				Miocene	
					23
		Paleogene		Oligocene	
				Eocene	
				Paleocene	
					65
	Mesozioc	Cretaceous			
		Jurassic			
		Triassic			
					250
	Paleozoic	Permian			
		Carooniferous	Pennsylvanian		
			Mississippian		
		Devonian			
		Silurian			
		Ordovician			
		Cambrian			
					540
Precambrian		Proterozoic			
					2500
		Archean			
					3800
		Hadean			
					4600

107

Tools of the Trade

Rock hammer, pick version
An essential field tool, used to crack open rocks to expose fresh surfaces and to pry crystals out of crevices.

Rock hammer, flat chisel version
This style of rock hammer is used for splitting shales and other layered rocks and is essential for exposing new fossil material.

Rock hammers

Digital camera
Essential both for documenting field sites and for photographing specimens in the laboratory.

Altimeter
Used to determine altitude; particularly useful in hilly or mountainous terrain.

Zodiac boat

Dynamite
Used to blast away overburden strata covering key fossil vertebrate deposits. Must be used with extreme care.

Point counter
Used to keep track of the number of points of a particular item (mineral type, fossil) under study. Typically each point will be the mineral present on a grid pattern placed over a polished rock slab or thin section.

Chisel
Used in combination with rock hammers to split rocks. The best ones have plastic-coated handles.

Brunton
The geologist's transit compass, used to measure the attitude of strata.

Small boat/Zodiac
Used to access otherwise inaccessible sites on islands and remote shorelines.

Secci (pronounced *SEK*-ee) disk
Used to assess light transmission through fresh or marine water.

Vernier caliper
Used to make precise measurements of fossils and mineral/sediment grains.

Vernier calipers

Hand lens, 10x and 20x
Used to identify minerals and fossils in the field.

Binocular microscope
Used for close analysis of samples in the laboratory.

Contact goniometer
Used to measure the angles between crystal faces in euhedral (well-formed) crystals.

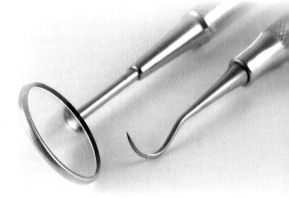

Dental tools

Dental tools
Used to remove matrix and better expose fossil specimens.

Tape measure
Used for measuring strata and other types of rock bodies; should be marked in both metric and British systems.

Field microscope
Used for close inspection of sediment grains and microbes in the field; carries its own power source.

Field notebook
Essential for recording field data about specimens, local geology, and field methods.

Covered clipboard
Used to protect maps and field notes from inclement weather.

Airless jackhammer
Used to break large slabs of rock and to fracture particularly resistant outcrops.

Rapidograph pen and ink
Even in an age of computer graphics, still useful for marking geologic maps and sketching fossils and minerals.

Rapidograph ink pen

Magnetic-tipped pen
Convenient way to test for rock's magnetic properties in the field.

Field sample marker
Indelible-ink pen used to label rock samples.

Portable rock saw
Gasoline-powered saw used to slice off critical slabs of rock in the field.

Flume tank
Used to study the properties of sediment in motion in a stream of recirculated water; also used to study the passive filter-feeding properties of ancient reef-building organisms such as archeocyaths and sponges.

Compact binoculars
Helpful in the field to view the rocks on the other side of the canyon; also known as the "Swiss rock hammer."

Binoculars

Topographic map
An ideal base map on which to place geological map information.

Stella software
A critical piece of software used to analyze feedback mechanisms by means of graphically created differential equations; essential for studying geochemical systems.

Topographic map

Image processing and drafting software
Essential for labeling photographs, giving them scale bars, and creating maps and stratigraphic sections.

Rock crusher
Used to crush rock samples for acid dissolution and/or geochemical analysis.

Stereoscopic air photos
Used to give a three-dimensional image of the land surface; extremely useful for geological mapping work. May be used with a pocket stereo viewer.

Gravity corer
Used onboard a ship or boat to take a shallow sediment sample from the sediment-water interface at the bottom of a body of water.

Sample bags, labeled
For protection, storage, and transport of samples from the field to the laboratory. Burlap cloth is best for rock samples; plastic is essential for wet samples.

Folding shovel
Used to remove accumulated soil and other debris from geologic sites.

Whisk broom
Useful to clean dust and dirt off shale and other rock surfaces in the field.

Paleomagnetic sample drill
A gasoline powered, chain saw–like device used to drill cylindrical plugs from sedimentary and igneous rocks for paleomagnetic analysis.

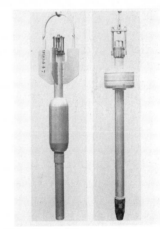

Phlegel's gravity corer

Jacob's staff
A yardsticklike staff with an inclinometer needle mounted at its top, used to measure stratigraphic thicknesses of inclined strata.

Tools of the Trade (continued)

GPS
A portable Global Positioning System device that uses satellite fixes to establish precise field position coordinates.

Petrographic microscope

Petrographic microscope
A specialized microscope that uses crossed polarizer plates to study mineral birefringence and other properties as seen in thin section.

Video camera for microscopes
Used to show petrographic microscope images to more than one person at a time; requires trinocular petrographic scope.

Scanning electron microscope
Used to greatly magnify and image the surface of geological materials using an electron beam generated by a tungsten filament.

Data loggers
Miniature portable devices used to record and store or transmit environmental data such as temperature and light intensity in remote locations.

Cathodoluminoscope
A dedicated petrographic microscope, electron gun, and vacuum chamber system used to detect the presence of quenchers and activators of cathodoluminescence in minerals as seen in thin section.

Submersible
A small submarine used to take geologists to remote field sites on the seafloor or at the bottom of deep lakes.

Robotic volcano explorer
A four-wheeled robotic probe used to carry cameras and sensors into volcanic fields where recent activity poses a threat to human life.

Electron microprobe
Used for nondestructive chemical analysis of geological materials.

Alpha-proton X-ray spectrometer
Used to determine the composition of planetary, moon, asteroid, and comet surfaces.

Moessbauer spectrometer
Used to analyze the oxidation state of iron in iron-bearing minerals.

A Moessbauer spectrometer and alpha-proton X-ray spectrometer in use on Mars

Thin section of rock

Thin-sectioning equipment
Used to cut billets, mount them to glass slides, and grind thin sections for final polishing on a lapidary wheel.

Transmission electron microscope
Used to greatly enlarge and produce an image of the ultrastructure of materials such as organic fossil wall layers that have been thinly sectioned for analysis.

Sediment sieves

Sediment sieves
Used to grade loose sediment samples into size categories.

Sieve shaker
Used to shake the sieves and sort the sediment sample quickly.

Mass spectrometer
Used to assess isotopic ratios in samples by means of measurement of mass-to-charge ratios of ions from the sample.

Core sample X-ray equipment
Used to pass powerful X-ray beams through sediment cores and other geological samples.

X-ray diffract meter
Used to determine the identity of minerals by analyzing their atomic structure.

Geiger counter
Used to detect radioactivity in rocks and minerals by measuring beta and alpha particle emissions.

Safety goggles or glasses
Essential to protect one's eyes from high-velocity rock chips.

Sledgehammer
Used for breaking large rock and for taking samples from tough rocks such as quartzites and eclogites. Safety glasses required!

Energy dispersive spectrometer
A device capable of discriminating between X-ray energies after a material has been bombarded by the electron beam used to image with the scanning electron microscope. This identifies the mineral composition of the materials.

Box corer
Used to take samples from shipboard of the sediment-water interface.

Soil corer
A hand tool used to take shallow soil samples for chemical or other analysis.

Ultraviolet lamp
Used to detect florescence in minerals and rocks.

Rock saw
Used to cut rocks with a diamond-studded cutting wheel; a variety of sizes are needed to handle most rock samples.

Safety goggles

Drilling rig
Used to drill and collect core samples of rocks, sediments, and soils. May be ship-based, land-based, or ice platform–based.

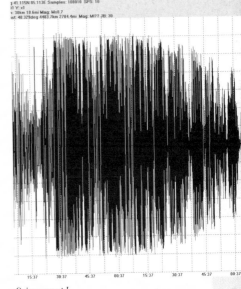

Seismograph

Seismograph
Used to detect and record seismic waves.

Field hat
Helps geologist keep cool and look stylish in the field.

Geiger counter

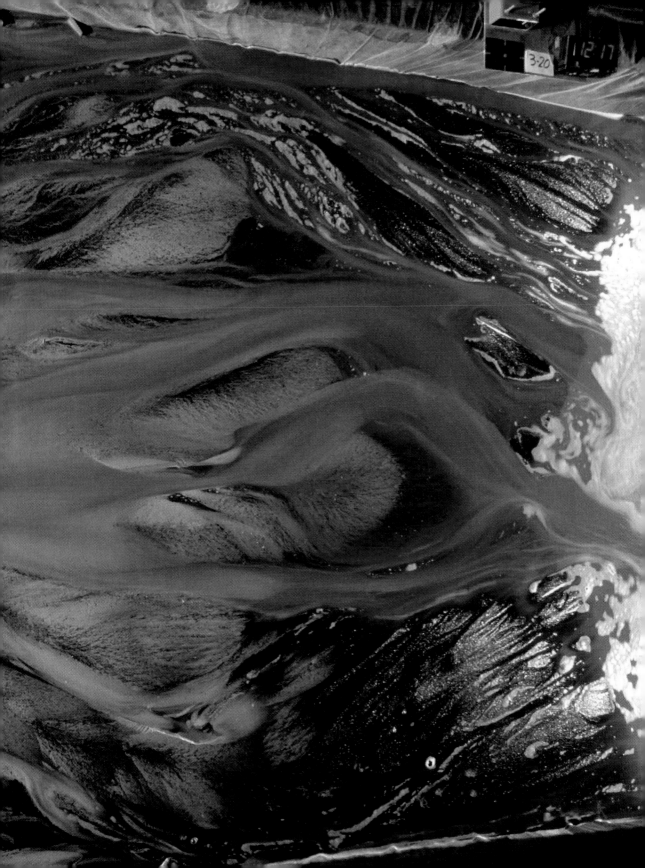

CHAPTER 7

GO WITH THE FLOW

Left: To test how sediment layers are laid down by river and ocean currents, researchers built a sand tank in which braided streams flow over sediment to a model sea in an experimental basin. Top: The clear water of a classic mountain landscape has become an increasingly scarce natural resource. Bottom: The orange tint in this polluted forest river is due to accumulation of metal oxides.

Uncontaminated water has been called the "new oil," a natural resource that requires careful stewardship if it is to remain widely available. Integral to this stewardship is study of the hydrologic cycle, which gives us information about the flow of water on the Earth's surface and in the relatively shallow subsurface. A subset of this study is the consideration of groundwater and its subsurface flow. The study of groundwater must rely on a variety of models to simulate its behavior, ordinarily hidden from view. Among these are mathematical models, computer models, and sand tank models. Mathematical models, enhanced by increasingly powerful computing resources, are based on the classic equations of Pierre-Simon, Marquis de Laplace (1749–1827), Henry Darcy (1803–58), and M. King Hubbert (1903–89), and are used to construct computer models that simulate groundwater flow. Yet models have the inherent limitation, as put by Herbert F. Wand and Mary P. Anderson, of attempting to solve "the problem of scaling down a field situation to the dimensions of a laboratory model." The available models of groundwater flow receive essential support from actual geophysical data such as that obtained by seismic refraction surveys.

The Hydrologic Cycle

Groundwater flow forms a significant part of the terrestrial hydrologic cycle. All freshwater bodies, lakes, glaciers, and the like represent only a drop in the bucket compared to the volume of water in the oceans. Nevertheless, this distilled or freshwater is an essential part of the hydrologic cycle, and it is absolutely crucial for the continuation of human life and myriad other life-forms. Water evaporated from the seas enters the atmosphere, where it is carried as water vapor—either invisibly or as water droplets (clouds) or as tiny ice crystals—that eventually rains down on both land and water. Large amounts of rain on land fall where moisture-laden air reaches a coastline and then is forced to higher altitude by a mountain range or front. As the air mass rises, it cools, and as it chills it can no longer hold the water as vapor, and rain or other forms of precipitation begin to fall. This cooling can be rapid and dramatic, as for instance the pelting hail that falls before thunderstorms in the Sierra Nevada in California and many other places, even in the middle of summer.

METEORIC WATER

Such precipitation, or meteoric water, penetrates the soil and (assuming that it is not intercepted by plant roots) enters the groundwater reservoir. Some time after the rain has been absorbed, the soil can be partitioned into three zones: the vadose zone (zone of aeration), the capillary zone, and the phreatic zone. Vadose water is water that passes through the zone of aeration but does not saturate it. The capillary zone, where water clings to soil and rock particles via capillary action (the ability of a substance to draw a liquid upward against the force of

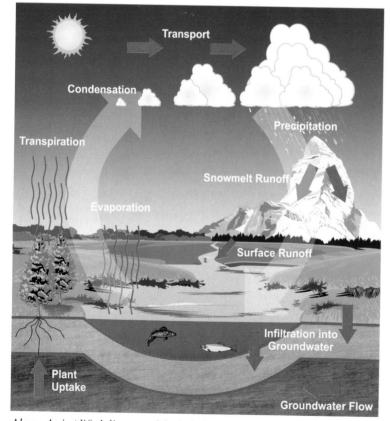

Above: A simplified diagram of the hydrologic cycle. Note that water vapor is returned to the atmosphere by evaporation from land, evaporation from bodies of water, or by transpiration by plants. Top left: A key element of the hydrologic cycle, precipitation in the form of rain, is a major source of freshwater.

gravity) forms a band (the capillary fringe) at the top of the phreatic zone. The term *phreatic* refers to groundwater able to flow freely in a zone of soil water saturation, and thus the phreatic zone is the saturated zone of soil below groundwater level. The depth from the surface of the soil to the groundwater surface can vary considerably and is dependent on the amount of rainfall associated with the particular local climate.

GROUNDWATER

A new area of geology is under development that is attempting to understand and quantify the relationship between the geological structure of an area and the associated groundwater flow velocity field over multiple scales. C. F. Tolman's influential 1937 text, *Ground Water*, presented his observation that "every geologic body has a geologic structure and a related hydrologic structure." This perspective continues today as a "permocentric" view of

Glacial Lake Hitchcock. Twenty thousand years ago, a thick ice sheet—as deep as two miles in most places—covered all of what is now New England.

geology, in which the thermal state of rocks, the chemistry, and the hydrology are related to the permeability distribution of any given rock body. Associated with this perspective is the concept of ideal versus nonideal aquifers. With the complexities of modeling as noted above, paleohydrologists in New Mexico and elsewhere are turning to ancient (and now dry) deposits that record evidence for ground-

water flow. In other words, they are using the "past as a key to the present." For example, accretionary nodules formed in deep soil deposits show asymmetric concretionary layers and thus evidence for groundwater flow, whereas nodules from ancient Lake Hitchcock in Hadley, Massachusetts, show more perfectly concentric laminations, indicating formation in more still-water (or stagnant) conditions.

A satellite image of the barren landscape between the Kunlun and Altun mountains in Western China clearly illustrates the striking shape of an alluvial fan deposit.

ALLUVIAL FANS, CALICHE, AND THE INITIATION OF SOIL FORMATION

The poorly sorted rocks and sand of alluvial fan deposits often form in semiarid environments and are thus liable to form soils with caliche horizons due to precipitation of calcium carbonate dissolved by infiltrating (and rapidly absorbed) rainwater. Well-developed caliche can form impenetrable, concretelike layers that prevent downward infiltration of water. Caliche is usually lighter in color than the soil in which it forms, and can be recognized in excavated soil by its color and resistance to crumbling.

Groundwater Contamination

Tracking plumes of contaminants can help protect drinking water. Locating and protecting sources of pure water can be a significant geological challenge. Once undesirable material has entered ground-the local groundwater table. Groundwater level (the water table), in spite of the name, is not always horizontal. Therefore the first questions to ask involve the gradient of the water table in the area of concern, and the Finally, strategies for stopping or containing the spill must be considered, and once a particular strategy is decided upon, it becomes the core of what is called the remediation plan.

LEAKING GAS
Consider the case of a gasoline (petrol) station with underground gas storage tanks. Tanks can rust through with time, and once breached, these tanks begin to leak their contents into the soil (a particularly lamentable loss in times of expensive gasoline). From the point of leakage, the gasoline will flow more or less straight downward to the inclined level (or gradient level) of local groundwater. Here, the contents of the spill will separate into two fractions. The upper fraction, immediately above groundwater level, will be a reservoir of free gasoline. Immediately below this, at the upper edge of local groundwater, will be a pool of dissolved gasoline components that have entered groundwater. Both the free gasoline (above groundwater level) and the dissolved gasoline components will begin to flow down-gradient (and to a minor extent up-gradient), following the slope of the local groundwater gradient. The dissolved gasoline

Above: Oil slick on water. Humans are highly sensitive to such hydrocarbon contamination in water supplies, which constitute a real threat to human— and other animal species'—health. Top left: An environmental technician tests water to determine its level of contamination.

water, all the tools of hydrology research come into play to track the plume of pollutants and to contain the contamination.

TRACKING THE SPILL
Once a contaminant plume has been identified, a number of questions must be addressed regarding the disposition of general direction of groundwater flow in this area. Flow is presumed to be down-gradient. Next, it is important to calculate the average velocity of contaminant movement since the time of release. This value can be used to predict the amount of time that will be required for the plume to reach an uncontaminated area.

Artesian wells contain waters at depth that are under hydrostatic pressure, and thus when tapped by a natural spring or an artificial well will flow to the surface on their own accord.

ARTESIAN WELL

An artesian well can form by tapping into a confined aquifer, a natural reservoir that is under hydrostatic pressure, a pressure that will cause water to flow to the surface (like a spring) when the aquifer is artificially breached. The pressure in an artesian well is maintained by a reservoir seal, a more or less impermeable layer of rock that prevents rapid flow of groundwater.

REMEDIATION

A hydrocarbon spill such as gasoline will dilute as it moves from the contamination source to disperse across the local water table. Nevertheless, we humans are dependent on pure sources of water, and our bodies have quite a low tolerance for hydrocarbon contaminants. Where drinking supplies are threatened, it may become imperative to pump contaminants out of the subsurface and destroy them in specially designed high-temperature furnaces installed on the site. In other situations, it may be sufficient to encourage the growth of soil microbes capable of metabolizing the contaminant hydrocarbons and rendering them inert.

components will flow/diffuse somewhat faster down-gradient than will the free gasoline.

How can one determine what is the direction of groundwater flow—in other words, the slope of the groundwater gradient? The flow of groundwater in a shallow aquifer generally tends to mimic the flow of water at the surface. Therefore, surface topography can be used to estimate the flow of groundwater even before subsurface work has begun. Subsurface investigations can be very expensive, especially if they require drilling, so careful assessment of surface topography and stream-flow patterns is essential for cost containment in any study of a contaminant plume.

An NRCS district conservationist and a local community college student assess the condition of a Pennsylvania river contaminated with iron oxide from a nearby mine. The mine-water remediation project will involve inducing the iron oxide to settle out—leaving behind cleaner water.

Well Construction

The ability to pull pure water from beneath the surface of the Earth has been a great boon to human existence. Civilization is greatly dependent on such supplies of clean water. Water wells come in a variety of types, generally depending on what they are drilled or dug into (rock or sand, for example) and the nature of the water source being tapped (confined or unconfined). Four primary types of wells can be constructed: dug wells, bored wells, driven wells, and drilled wells.

DUG WELLS

Dug wells come in two varieties, protected and unprotected. Both types are typically shallow wells. Unprotected dug wells,

the most primitive wells, are holes in the ground dug to the level of the local water table. Lowering and raising a bucket with a rope facilitates access to the water. Dug wells are obviously unsanitary should undesired items such as animals fall into them.

A more modern protected well has the well hole lined in its lower part with unmortared brick. This unmortared section extends down into the water-bearing formation. A gravel filter bed is placed at the bottom of the well. Above this is a curbing course of mortared brick surrounded by two

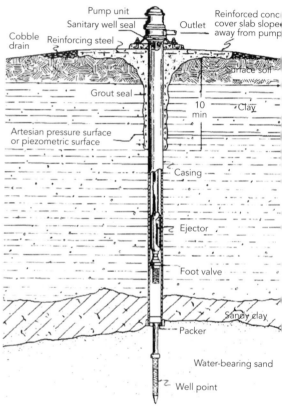

concentric layers of 6-inch-thick (15 cm) reinforced concrete. This curbing typically extends for 10 feet below the surface of the ground. A circular well cover seals the top of the well, the edges of which must extend 18 inches (45 cm) beyond the diameter of the well itself. Two holes penetrate the well cover: one in the center for installation of a pump, the other off

Left: An example of a dug water well. This 100-year-old well in Argentina was lined by hand with bricks. Above right: Diagram of a bored well with a driven well point. Note that the grout seal extends some distance below the wellhead. Top left: Masonry stonework surrounding a permanent water well.

center to hold a valve that provides access for chlorine. The pump hole should be lined by a sleeve that extends well above and below the top and bottom surfaces of the well cover. The pump cylinder (where water is drawn by the pump) should be situated so as to remain permanently submerged.

BORED AND DRIVEN WELLS

A bored well has characteristics similar to a dug well. Smaller-diameter bored wells may be lined by a metal cylinder (steel casing) driven into the bored hole, an arrangement that will work well as long as there is a filter screen at the bottom the cylinder. Driven wells are somewhat similar to small bored wells because they are lined with steel pipe. In this case, however, the pipe is driven into the ground to the level of the water table. The well point at the end of this pipe has pores through which the water may flow into the pipe. The well point should project entirely into the aquifer stratum such as a water-bearing sand. Once pounded in, the driven well is usually encased in a tube driven to a depth of 10 feet (3 m). This type of well, where the discharge pipe reaches the surface, has a 4-foot (1.25 m) square concrete platform around the well. The pump stand and the base must be in a single piece, and both an asphaltic seal and a grout seal connect the outside case to the concrete platform.

Locating water deeper within Earth's surface calls for heavy machinery, such as this water well drilling rig in West Virginia.

DRILLED WELLS

Dug, bored, and driven wells all tend to be fairly shallow, and therefore may be operated by hand pumps. Drilled wells, on the other hand, tend to be small diameter and much deeper. They are therefore typically operated by power pumps. To construct such a well, a small-diameter hole is bored into bedrock and casings are added to this uncased drill hole. The casing generally ends where the bedrock begins. A special grout seal is placed where the well casing intersects the upper bedrock surface to forestall seepage along the outside of the outer casing.

Wastewater and Runoff Water Treatment

In his 1856 book, *The Public Fountains of the City of Dijon*, Henry Darcy first elucidated his celebrated Darcy's law experiments that are now fundamental to our understanding of water discharge. Using his now famous sand column, an "apparatus designed to determine the law of flow of water through sand," Darcy was able to show experimentally that the volume discharge rate is directly proportional to the head (elevation) drop and to the cross-sectional area of the column, but inversely proportional to the length difference. Understanding the flow characteristics of water through sand is essential for water treatment because sand or other types of particulates can be useful as purifying filtration systems for water.

CISTERNS

In areas where groundwater flow may be insufficient for local needs, rainwater may need to be collected in cisterns. Cistern waters, however, have the decided disadvantage of being more subject to contamination (by bird droppings, soot, soil in gutters, and so on) than are groundwater supplies. Cistern water, being a product of direct runoff from precipitation, lacks the advantage of the natural filtration provided by passing through soil with its filtering characteristics and microbiota hungry for most types of dissolved organic chemicals. Nevertheless, a cistern with a sand filter can be an effective means of generating pure drinking water. A sand filter is placed above the cistern reservoir, which is typically buried beneath the surface and lined with an impermeable double layer of reinforced concrete. The sand filter, which resembles sandy soil profile in miniature, consists of a 4-inch (10 cm) layer of coarse gravel at the base, covered by 3 inches (7.5 cm) minimum of fine gravel, covered in turn by 3 inches (7.5 cm) minimum coarse sand, covered finally by 20 inches (50 cm) minimum of filter sand. A pyramid-shaped galvanized metal screen covers this artificial pile of stratified sediments. This sand filter system is enclosed in a cylinder formed by a double layer of reinforced concrete

Above: Water recovery by bucket from a cistern, a traditional water-collection method. Top left: Runoff from saturated soil. Rock outcroppings nearby suggest that the soil is thin here and has limited water storage capacity.

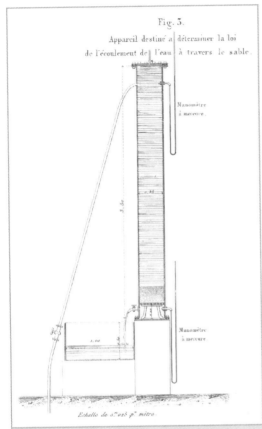

An 1856 illustration of the Darcy flow tube.

DARCY FLOW TUBE

This archival image from Henry Darcy's *The Public Fountains of the City of Dijon* shows his experimental sand column. Note the mercury manometer attached to the side of the sand column. This famous apparatus, and the calculations based upon it, are considered by many to be a criterion example of the proper approach to experimental quantification in the earth sciences.

valve for times of high water flow. The pump cylinder projects into the cistern as well, and should remain permanently submerged. Cisterns must be carefully situated so as to not be susceptible to contamination by cesspool wastes. Proper cistern development will be very desirable for the development of permaculture projects in Brazil and other parts of the world.

ADVANCED WATER TREATMENT

Humans produce large amounts of organic waste, and much of this is contaminated with coliform bacteria derived from the human gut system. Chlorine can be used to eliminate these bacteria, but it takes one ounce of

water. Under development are new systems of water treatment that go beyond the massive bacterial consumption of human waste at (typically) ugly and foul-smelling waste treatment facilities. Educational and research organization Ocean Arks International has developed an intriguing system of water treatment that involves seven different water tanks connected in concert. Untreated sewage from a cesspool lagoon goes into the first tank and is passed to each successive tank in turn. Each tank contains its own particular microbial community that is specialized to absorb nutrients from the progressively diminishing waste stream. Incredibly clean water flows from the seventh tank, microbially purified of organic pollutants.

Circular wastewater treatment pools. Bacteria and other microbes consume waste nutrients, purifying the water before release into a local river.

(or in some cases, steel sheets), and is penetrated at the top by a downspout that acts as the rainwater inlet and at its base by a discharge tube that delivers water to the cistern. Penetrating the side of the sand filter enclosure near the top is an overflow

hypochlorite (containing 70-percent-available chlorine) to disinfect 100 gallons of water, and the chlorine breakdown products can introduce undesired by-products (toxic to goldfish and many other organisms) into the drinking

Karst and Caves

Karst topography develops in areas with a humid climate and layers of limestone in the ground. Karst represents an interesting combination of erosion by acidic groundwaters and mass wasting by sinkhole and cave collapse. In a mature karst landscape, most of the limestone or dolomite bedrock has eroded away, leaving sharp-spired remnants of the walls of cave systems that now project into the sky.

KARST HYDROLOGY

Rainwater is naturally acidic, for it forms carbonic acid as a result of picking up dissolved carbon dioxide during its fall through the atmosphere. When this slightly acid rainwater falls on a limestone or dolomite terrane, the bedrock begins to dissolve because of a chemical reaction between the calcium carbonate of the rock and the carbonic acid of the meteoric water. Beginning with fissures and fractures in the rock, the carbonic acid (and in some situations, sulfuric acid) in the water attacks the carbonate rock, forming underground networks of channels and caves that taken together constitute an enormous subterranean drainage system.

KARST AND CROPS

The astonishing underground drainage systems of karst regions can pose serious problems for agriculture in these areas.

Soils in karst areas tend to be good to excellent for plant growth, and rainfall is typically generous (otherwise the karst would not have had a chance

Above: A view of the Pool of Peace at Howe Caverns in New York State. Limestone stratification is visible in the cave opening. Below: Karst cliff in Herault, France, features abundant caves, favored shelter for both wild animals and early human inhabitants. Top left: Terrace karst on Navassa Island off Haiti shows the irregular surface of the carbonate rock where dissolution has occurred.

Calcium carbonate–saturated stalactites, forms of speleothems, hang from the ceiling of Punka Cave in the Moravian karst area of the Czech Republic. Stalagmites, another form of speleothem, would rise from the floor.

to form), but the draining of runoff water into the subsurface channel system is so rapid that it can leave the surface soils parched. Furthermore, contamination is a serious problem in karst areas, because groundwater flow is intercepted by the underground drainage channel system and is denied the natural filtration and purification that occurs by slow passage through soils and sand deposits. An additional complication is that karst caverns and sinkholes are frequently used as convenient dumping grounds for a variety of wastes, further exacerbating the problem of lack of natural filtration. Finally, heavy farm equipment has on occasion been swallowed up by collapsing sinkholes, a double problem because of both the loss of this expensive machinery and the fact that it introduces organic contaminants such as gasoline and motor oil into the subsurface flow. Extra caution is therefore required when managing the hydrology of karst regions.

CAVERNS

Caves in karst areas constitute some of the natural wonders of the world. Formed by the slow dissolution of limestone by acidic waters, the caves ironically capture back some of their lost calcium carbonate in the form of layered rock deposits, some with fantastic shapes, known as speleothems. The best-known speleothems are perhaps the stalactites that hang from the roof of caves. These drip water supersaturated with calcium carbonate. The drops splash to the cave floor to form the stalagmites that grow upward. Caves often host underground lakes, and these lakes can sustain unique species of fish, arthropods, and other types of animals that have become blind due to millions of generations below the surface away from sunlight. Cave systems can even host underground streams, such as the River Styx in the 10-million-year-old Howe Caverns of New York State, where one can take a boat ride.

This eighteenth-century ink on silk Chinese landscape was clearly inspired by a karst landscape.

KARST IN CHINA

Although many classic depictions of the Chinese landscape look like fanciful flights of the imagination, they are not examples of artistic license. This painting of a karst landscape in China shows the classic spires of mature karst topography. Each spire was once part of the wall of a cave or underground channel. As the caves collapsed via sinkholes and other mass wasting phenomena, the more resistant cave walls were left standing as the cavern-riddled bedrock collapsed around them. Karst spires should not be confused with stalagmites, which only grow upward from the floors of caves.

Ocean Circulation

In terms of sheer volume of water transported, no part of Earth's hydrological system matches the amount of water carried by ocean currents. These huge currents control the temperature of continents, patterns of rainfall, and many other important features of Earth's climate system. Ocean currents may be divided into two main types: surface currents and currents at depth. Surface currents are driven by circulation cells and associated winds, and these currents follow curved trajectories because of the Coriolis effect (the tendency for any moving body on or above Earth's surface to drift sideways from its course because of Earth's rotation) and because they often tend to parallel the margins of the roughly circular major ocean basins. Deep-water currents are generally driven by density contrasts (due to high salinity or low temperature) between the waters that flow at depth and the less dense waters above them.

EARLY RESEARCH

Beneath the surface of the Mediterranean flows a warm and high-salinity current toward the Strait of Gibraltar. This current flows out of the strait and on to the eastern Atlantic continental slope, and can be traced for thousands of miles from its source at the strait. Ancient Phoenician navigators were aware of this flow, and would drop umbrella-shaped submarine sails down to the hypersaline current. This would allow them to power their wooden ships westward,

Above: Famed explorer of the New World Juan Ponce de León is credited with the discovery of the Gulf Stream. Left: Benjamin Franklin's chart of the Gulf Stream, which greatly facilitated water travel from Europe to the United States. Top left: The oceans, which cover three-quarters of the planet's surface, transport the greatest volume of water in the hydrological system.

in some cases against prevailing winds. The discovery of the Gulf Stream, attributed to Ponce de León in 1513, was a great boon to transatlantic navigation, for, along with its equatorial countercurrent, this current system provided natural assistance for both the outward journey from Europe to North America (via a southern route) and help for the return journey (via the Gulf Stream proper) back to northern Europe.

MAJOR OCEAN CURRENTS

It is now known that major oceans on Earth with a significant north-south dimension will, at any time in Earth history, develop two major current systems joined at the equator by a westward-flowing equatorial current. In the Northern Hemisphere, the current system will flow clockwise. For example, the Gulf Stream is the northwestern segment of this circulation system in the North Atlantic. In the Southern Hemisphere, the flow will be counterclockwise, forming something like a mirror image of the circulation north of the equator. Major areas of upwelling, which support great marine bioproductivity, such as the anchovy fisheries off the west coast of South America, are associated with the areas on the eastern sides of great oceans where the current of the major circulation cells are moving offshore. This offshore movement causes cool, nutrient-rich waters to move up to the surface in nearshore areas.

Satellite view of the Gulf Stream region off the northeastern coast of Canada and the United States; note the lighter color of the warmer Gulf Stream seawater. A ring current is visible near the center of the image.

RING CURRENTS

Where a warm current such as the Gulf Stream flows into a cooler region, such as the North Atlantic, a pronounced thermal contrast between the current waters and the surrounding waters will develop. Both the temperature and the color of the waters can be easily detected from the deck of a ship. The Gulf Stream is thus part of a tremendous transoceanic conveyor belt that makes the winter climate of northern Europe much milder than it would be otherwise. Somewhat like the meander bends in a river course, the Gulf Stream can develop meander loops. On occasion, these meander loops will pinch off into a ring current that becomes separated from the mainstream of the Gulf Stream. There is a tragic element to ring currents, for the warm-water life-forms that the warm ring current waters support will slowly perish as the ring current slowly decelerates and begins to mix with the cooler surrounding waters.

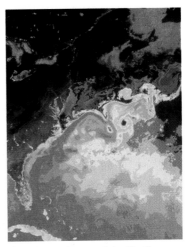

A computer image of the North Atlantic, showing ring currents with warm cores off the eastern seaboard of the United States.

CHAPTER 8

GLACIATIONS THROUGH TIME

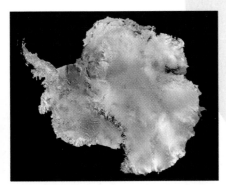

Left: The stunning beauty of a glacial waterfall on Mendenhall Glacier in Juneau, Alaska. Top: Antarctica as seen by satellite. Bottom: A glacier in Greenland. The polar ice caps of Antarctica and Greenland attest to the fact that the Earth is still in a state of glaciation.

Repeated ice ages have greatly influenced Earth's climate over the past three billion years. Some of these glaciations seem to have been triggered by biotically driven changes to the composition of the atmosphere, although tectonically driven uplift, Earth's orbital variations, and rock weathering surely also play roles in the onset of glacial conditions. The challenge here is to understand how the Earth gets out of an ice age once it has fallen into it. There have been a number of glaciations in Earth history, ranging in age from more than two billion years ago to only a few thousand years ago. In spite of evidence for global warming, we are still in a glacial period, as seen by the ice caps covering Greenland and Antarctica. The primary scientific questions regarding study of ice ages include the following. Why do glaciations occur? Why do they occur when they occur? And, as just noted, what causes glaciations to end? Scientists generally split into one of two camps when considering the first question. The first camp sees glaciation as primarily controlled by physical and chemical processes operating on Earth. The second camp views glaciation and deglaciation as being driven primarily by the activity of the biosphere.

Early Glaciations

Ever since amateur geologist Peter Dobson's interpretation of boulders in the Connecticut Valley as glacial erratics, Swiss engineer Ignace Venetz's exposition in 1833 of the classic theory of glaciation, and Swiss-born American zoologist, glaciologist, and geologist Louis Agassiz's presidential address at the 1837 meeting of the Swiss Society of Natural Sciences of Neuchâtel promoting the concept of *die Eiszeit* (the Ice Age), ice ages have become a widely accepted fact of Earth history. Dobson, who like a glacial erratic boulder himself, emigrated from Britain by illegally hiding in a hogshead barrel rolled up onto an America-bound ship, wrote the seminal article "Remarks on Bowlders" in 1825 for the *American Journal of Science*. It subsequently has become clear that there have been multiple episodes of glaciation on Earth. The severity of some of these glacial intervals, however, has been less well known until more recently. With the development of White Earth, or Snowball Earth, theory, it has become clear that particular intervals during pre-Phanerozoic Earth history were marked by episodes of severe glaciation.

Above: Louis Agassiz (1807–73) proposed the existence of a past ice age in 1837. Although his idea was initially met with skepticism, the concept of die Eiszeit was soon widely accepted by the scientific community. Top left: "Glacier de Zermatt," an illustration in Agassiz's Études sur les glaciers, *published in 1840.*

THE FIRST GLACIATIONS

The earliest glaciations seem to be associated with the early expansion of photosynthesis and the generation of oxygen in the early atmosphere. Corneille Jean Koene and J. J. Ebelmen's mid-nineteenth-century research on the carbon cycle demonstrated that rock weathering and plants could control the levels of carbon dioxide in the atmosphere by absorbing carbon dioxide by their breakdown or growth, respectively. Carbon dioxide is of course a critical greenhouse gas. The positions of tectonic plates may also play critical roles. The first-known glaciation period, the Huronian glaciation (2.2 to 2.4 billion years ago), can be understood from the perspective of the Lomagundi-Jatuli carbon isotopic event.

LOMAGUNDI-JATULI CARBON ISOTOPIC EVENT

The Huronian glaciation seems to be associated with radical climate change and the most massive perturbation of the global carbon cycle in Earth history. The Lomagundi-Jatuli event (2.0 to 2.4 billion years ago) is a time when isotope ratios of carbon go wildly out of control, for reasons that are leaving geologists scratching their heads. The first red beds (oxidized sediment layers) appear at this time, and it is thought that these represent the earliest recorded beginnings of an oxygen-rich atmosphere. Thus, biochemical oxidation and reduction, known as redox, and changes in the biogeochemical reservoirs of oxided carbon

(carbon dioxide) and reduced carbon (methane) seem to be involved here. Because both carbon dioxide and methane are potent greenhouse gases (with methane the more potent of the two by a factor of 20), it seems reasonable to suppose that both the first glaciation and the Lomagundi-Jatuli event were in some way caused by biological utilization of carbon.

THE GOWGANDA FORMATION

Further evidence for a close association between changes in redox geochemistry and the earliest glaciations comes from the Huronian Supergroup in Ontario, Canada. This body of rock, deposited between 2.2 and 2.4 billion years ago, contains multiple glacial units that attest to severe ice-age conditions. The youngest of these is called the Gowganda Formation. The 4,600-foot-thick (1,400 m) Gowganda Formation

The Gowganda Formation, part of the Huronian Supergroup, contains all of the features one would expect to find in a glacial deposit, including water-layered deposits and till. What makes it remarkable is its age. At 2.3 billion years old, it is one of the oldest-known glacial deposits on Earth.

is divided into a lower unit of tillite (stratified glacial till) called the Coleman Member and overlying shale called the Firstbrook Member. Some shale is present as well in the Coleman Member. Takemaru Hirai and his colleagues at the University of Tokyo have shown that the Coleman shales are green

colored (reduced), but that the shales at the top of the Coleman Member are oxidized red, due to high hematite content. Joseph Kirschvink at California Institute of Technology in Pasadena argues that massive amounts of manganese- and iron-rich sediments were deposited at the end of the Gowganda glaciation.

MILANKOVITCH CYCLICITY THEORY

Milankovitch cyclicity theory attributes 100,000-year and 40,000-year cycles in glacier expansion and contraction to variations in the Earth's axial tilt and orbit parameters. These cycles are known to influence the rates of Earth processes. For example, strata deposited in Mesozoic basins along the coast of eastern North America show evidence of ice sheets having grown and melted over short intervals. These changes are attributed to cyclic wobbles in Earth's rotation and orbit, otherwise known as Milankovitch cycles.

Milutin Milankovitch, the Serbian geologist who theorized that climatic change is affected by changes in the tilt of the Earth's axis. According to Milankovitch theory, higher tilts bring about more extreme seasonal changes, while lower tilts account for milder ones.

Snowball Earth

Snowball Earth, or White Earth, theory holds that during the most severe glaciations, the oceans essentially froze over and the atmosphere became so cold that carbon dioxide itself might have frozen out of the atmosphere at the poles, further dropping the Earth into a severe glaciation, or what has been called icehouse climatic conditions. The Snowball Earth theory is controversial, with many Earth scientists denying that the oceans could have ever frozen over solid.

TRIGGERING THE SNOWBALL

Did the Earth's surface ever freeze solid, or were there always polynyas, or cracks in the ice? If so, once frozen, how did Earth recover from the deep freeze? And how could Earth fall into such a severe glaciation in the first place? Although the phrase "Snowball Earth" has been applied to any major glaciation in Earth history (for "major," read any glaciation considerably more severe than the Pleistocene ice age), the term primarily refers to a series of glaciations that occurred in Late Proterozoic time, between about 750 to 600 million years ago. The exact number of glaciations is currently under debate, due to difficulty in arriving at precise dates for the various tillites of this time interval. We know that there were at least two and more likely three primary phases of glaciation. Of the two main ice advances, the first is called the Sturtian glaciation and the second is called the Marinoan glaciation.

ICE AT SEA LEVEL

The Late Proterozoic glaciations are notable for one particular feature that evidently attests to their severity. Paleomagnetic measurements of glacial strata

Above: According to the Snowball Earth theory, a large amount of equatorial land and decreased levels of atmospheric carbon dioxide led to a massive, planet-wide cooling around 590 million years ago. As glaciation ensued, the sea was covered by kilometer-thick ice that lasted for millions of years. Top left: A frozen-over waterfall in present-day Antarctica. Is it possible that the Earth was once a vast planet of icy seas?

from these glaciations (most notably from deposits in Australia) convincingly demonstrate that ice sheets formed relatively close to sea level right at the ancient equator. This astonishing result was hard for many geologists to accept, but the result has been duplicated numerous times, and the data are quite convincing. This is the single strongest piece of evidence favoring the Snowball Earth theory, the idea being that if ice sheets can form at sea level at the equator, then it is possible that the entire Earth was glaciated and that the ocean was covered with an icy crust.

CAP CARBONATES

A remarkable and equally puzzling feature is associated with the glacial strata from both the Sturtian and Marinoan glaciations. This is the presence of a distinctive layer of dolomite (or rarely, a layer of limestone) that was deposited immediately on top of the glacial strata. This rock layer is called the cap carbonate because of its carbonate composition and because

it forms a "cap" over the presumed tillites. The texture of the dolomitic sediments in these cap carbonates is unusual, and sometimes shows the development of crystal fans and other features of precipitation from seawater that is supersaturated with calcium carbonate. Such features are usually associated with calcium carbonate that is deposited in warm waters. Such thick and extensive layers of carbonate precipitated in this way are rare to unknown elsewhere in the geological record. The cap carbonates appear to record an interval of extremely hot climate, perhaps hotter than at any time in Earth

history since. So here we are left with an astounding paradox: the evidence for the coldest time in Earth history is juxtaposed against evidence for the worst episode of global warming ever known. All this provides a rather sobering testimony to our planet's potential for sudden climate change.

Pink-hued cap carbonate found in China's Gobi Desert.

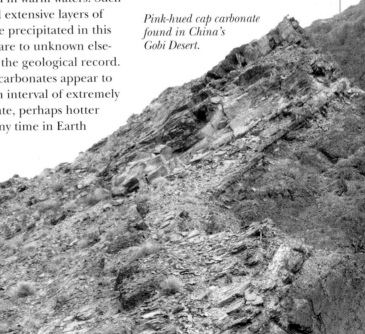

Tectonics and Recent Glaciations

The most recent glaciations are hypothesized to have been triggered by tectonic changes such as uplift of the Tibetan Plateau and closure of the Isthmus of Panama. The Earth is currently in a glacial-interglacial cycle. Has human alteration of atmospheric composition ended this cycle, or are we still in it?

THE HIMALAYAN CONNECTION

Beginning about 55 million years ago, tectonic uplift of the Tibetan Plateau and the Himalayas is thought by Maureen Raymo of Boston University and others to have cooled global climate to such

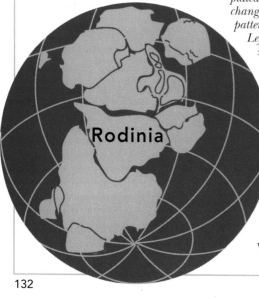

Above: A satellite image of the Tibetan Plateau, the largest and highest plateau in the world. Glaciation in this region directly influences climate change, and scientists fear that global warming that threatens to disrupt this pattern could have serious consequences for hundreds of millions of people. Left: Many scientists associate the tectonic uplift that broke apart the supercontinent Rodinia with the Snowball Earth glaciation. Top left: The Himalaya mountain range is believed to have played a large part in the onset of glacial conditions around the world.

an extent that it triggered the ice age cycles that began two million years ago. The proposed mechanism of this cooling is as follows: As the Tibetan Plateau was uplifted high above the land areas that surround it, pools of cold air began to form above the plateau. Cooler than the air masses over the land surrounding the plateau, this cold air, being more dense, began to flow downward and chill areas nearby. This chilling effect extended as far as North

America and even Europe, causing global temperatures to gradually decline until they triggered the onset of glacial conditions.

TIBETAN GLACIATION

Glaciation has also influenced the recent climate of the Tibetan Plateau itself, and the fluvio-glacial system of this region provides freshwater for perhaps a half billion people in Central Asia. Lewis Owen of the Department of Earth Science at the University of California at Riverside argues that alterations to this system, via global warming or other factors, could have profound political and economic consequences for hundreds of millions of people.

MOUNTAIN RANGES AND GLACIATION

Both locally (Tibetan Plateau) and globally, the orography (the physical geography of mountains and mountain ranges) of the Himalayas continues to play a key role in influencing the onset of glacial conditions. Perhaps not coincidentally, the worst

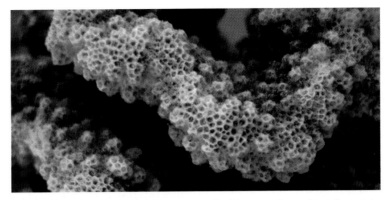

Oceanic cooling is detrimental to the growth of hermatypic corals and can trigger massive aquatic absorption of carbon dioxide.

glaciations on record (Snowball Earth glaciations of the Late Proterozoic) are also associated with an episode of tectonic uplift associated with the breakup of the supercontinent Rodinia. There is a difference, however, between the breakup of a supercontinent and the formation of a plateau such as the Tibetan Plateau. The former is the result of tectonic tension and fracturing, whereas the latter results from tectonic compression due to subduction and continent-to-continent collision. Nevertheless, both situations lead to the uplift of large masses of rock. In the case

of plateau formation, granitic rocks are thrust on top of other granitic rocks, and because of density relationships with the mantle (isostacy), great elevation increases ensue. In the case of continental rifting, uplift also occurs because of thermal uplift of granitic rocks along the rift scar due to upwelling of hot basaltic magma from the mantle. In both cases, fresh metamorphic rocks are exposed to weathering. The Urey reaction subsequently goes into operation on a massive scale, absorbing vast amounts of the greenhouse-gas carbon dioxide from the atmosphere. This carbon dioxide withdrawn from the atmosphere eventually ends up in the ocean as calcium carbonate sediment, sequestered in the rock record as coral reef limestone and other types of carbonate deposit. This part of the carbon cycle can act as a negative feedback on global cooling, for if the oceans become too cold, the growth of hermatypic (carbonate mound-forming) corals is drastically inhibited.

THE NORTH ATLANTIC HEAT CONVEYOR BELT

Waters of the Gulf Stream today cool as they reach the waters north of Europe, sink, and carry cold water toward Antarctica at depth in what has been called the great transoceanic conveyor belt. Evaporation along the Gulf Stream part of the conveyor increases the salinity and adds to the high density of the chilled North Atlantic water. In the Pliocene epoch, the Gulf Stream waters turned left instead of descending and flowed into the North Polar region. This led to pronounced warming of air masses in the polar region. Thus, ocean currents can play a significant role in global heat transfer.

Glacial Landforms

Glacial landforms, primarily seen today on continents of the Northern Hemisphere that experienced recent glaciation, include such features as moraines (accumulations of boulders, stones, or other debris carried and deposited by a glacier), drumlins (elongated ridges of glacial drift), and eskers (sinuous gravel ridges formed by subglacial streams). Other features such as frost wedges and patterned ground can also be directly

Above: Drowned drumlins, such as this one in Clew's Bay, Ireland, are submerged in the water and appear as a cluster of islands in the sea. Left: A glacial moraine found in Callaghan Valley, British Columbia. Top left: Glaciers leave distinctive landforms in their wake, such as this fjord.

observed as primary glacial features. These features have taken on added interest, as attempts are under way to recognize them on the Martian surface from spacecraft imagery.

GLACIAL LANDSCAPE

As a glacier retreats, either during an interval of interglaciation or at the end of an ice age, it leaves behind chaotic heaps of sediment in a variety of forms. This glacial debris tends to be unsorted, which means that a large variety of sediment grain sizes and densities are all

mixed together. This mixture of materials ranging in size from boulders to clay particles is called till when its glacial origin is certain. If the glacial origin of such material is not certain, it is then referred to as diamictite. Till that has been cemented to stone is called tillite.

MORAINES

Perhaps the most prominent features of a glacial landscape are moraines. These linear to curved ridges of till stand out in the lowland areas of a glacially sculpted landscape. Moraines form as accumulations of sediment on the edges of the glacier ice. But rather than merely representing material that is pushed along, as soil might be pushed by a plow shovel, moraines represent places where glacier ice reached a sort of dynamic equilibrium with local (and not so local) climate. In other words, the moraine that forms at the toe of a glacier (called the terminal moraine) marks the place where the rate of melting of the ice was more or less matched by the influx of new ice from newer ice formed by precipitation "upstream" on the glacier ice. Marginal moraines are moraines formed on the lateral edges of the glacier stream. They are particularly well developed in valleys that host glacier flow. Such valleys develop, as a result of glacial erosion, a characteristic U-shape. This distinctive profile distinguishes glacial valleys from the V-shaped valleys typical of stream and river erosion.

DRUMLINS

Drumlins are elongate or elliptical hills that are also characteristic of glaciated terrains. Aligned lengthwise with the direction of glacier flow, drumlins are formed by bedrock obstructions at the base of glaciers that force the glacial ice to release its entrained load of unsorted sediment. This sediment piles up into a mound of till in the vicinity of the bedrock prominence, protecting it from further erosion by the

ice and causing more sediment to accrete as the ice is induced to melt by pressure solution. Arrays of parallel drumlins, such as those encountered in New England, can give a distinctive look to the post-glacial landscape.

marks—crescent-shaped impact scars formed by violent skipping ahead of boulders right at the base of the glacier that undergo a sudden tension-and-release motion as the glacier that bears them surges forward.

This exposed bedrock surface in Glacier National Park, Montana, has been both polished and scratched by glacial movement. The scratches, called glacial striae, are parallel to the direction of the glacial flow.

STRIATED PAVEMENTS

The rock-grinding action of glacial ice is significantly amplified by the rocks entrained in the glacial ice. Like a use-worn piece of sandpaper, chunks of rock on the base of a glacier can both polish a bedrock surface and score it with linear scratches. Such parallel scratches are called glacial striae, and, like drumlins, they tend to run parallel to the direction of ice flow. Glacially polished bedrock surfaces can also develop chatter-

PATTERNED GROUND

Patterned ground includes a variety of stone ridges, stone networks, and stone polygons formed by frost disturbance of regolith. Delicate patterns that seem to ornament post-glacial landscapes can be formed in this way. There appears to be an element of natural self-organization in these patterned-ground polygonal networks of rock. Perhaps some of these patterns inspired early human megalith constructions.

Glaciation and Regulation of Climate

I t has been hypothesized that organisms modulate global climate by the summation of their metabolic activities (Gaia theory). It has also been hypothesized that life processes are capable of throwing the Earth climate system severely out of balance. How then should we view glaciations? A necessary course correction or a system in crisis? Do these theories over- or underestimate the influence of life on geology and climate?

GAIA AND GLACIATION

Gaia theory, which holds that organisms themselves are responsible for regulation of global climate over long stretches of geologic time, also claims that the Earth remains comfortably habitable because of robust biogeochemical regulation by the biosphere. Scientists who study Earth's geochemical systems have vigorously debated this perspective. Some scientists voice support by pointing out that during the last few million years, a time of glaciation in the Northern Hemisphere, the critical mid-latitudes of the Earth (containing 70 percent of the surface) have fluctuated only within about five degrees during this time. Others have argued that extreme glaciations pose the specter of the breakdown of any biotic global climate regulation system. James Lovelock, originator of the Gaia theory, argues that "this is not feeble

Above: Scientists at Vostok, Antarctica, use ice cores to study the history of Earth's climate. By drilling into the Antarctic ice, climatologists are able to extract ice cores that contain air bubbles that have been trapped in the ice by falling snow. Left: Vostok researchers pose with five ice cores taken from the Antarctic ice sheet. Top left: Satellite imager shows El Niño fires in California releasing greenhouse gases into the atmosphere.

regulation," but rather a maintenance of comfortable habitability that is comparable to the thermal regulation and homeostasis in our own bodies.

METHANE AND CARBON DIOXIDE

The Vostok ice cores have provided us with a spectacularly detailed record of climate and atmospheric composition over the past 400,000 years. These ice cores allow determination of ancient atmospheric composition by means of analysis of small bubbles of old air trapped in the ice. This remarkable data set has several features that are of unique importance for our understanding of glaciation. First, the atmospheric levels of carbon dioxide seem to track changes in global temperature more or less exactly. Second, the recorded levels of atmospheric methane over the same period track the fluctuations in both carbon dioxide level and global temperature just as exactly. What does this all mean?

GREENHOUSE GASES

These lockstep changes in the levels of two main greenhouse gases and global temperature imply either that global temperature is very tightly coupled to levels of greenhouse gases, or that increases in global temperature lead to releases of greenhouse gases. If the former, then climate forcing by greenhouse gases exercises a direct and more or less immediate control on atmospheric

Planetary reflectivity, or albedo, is a major factor in the regulation of the Earth's climate. The higher the Earth's reflectivity, the less heat the planet retains, thus leading to cooler temperatures. Conversely, low planetary reflectivity causes the Earth to absorb heat, leading to increased warming trends.

GLACIATION AND THE POTENTIAL FOR EXTREME CLIMATE

Some decades ago, there was a paleoclimate hypothesis in vogue called the "ice blitz theory." This concept posited that the onset of glaciation was breathtakingly rapid; with the onset of glacial conditions occurring in as little time as a single summer when the winter snows fail to melt, and the sudden increase in summer albedo (planetary reflectivity) drops the world into a glaciation. Most climate scientists today would consider this scenario to be unrealistically extreme. Nevertheless, Anthony Prave at the University of St. Andrews in Scotland notes that the "geologic record shows unequivocally that periods of oscillating climatic extremes are an integral component of how the Earth system has evolved and functioned." Prave continued, in discussing Snowball Earth, to say, "it should come as no surprise that humankind might, will, be faced with having to deal with another such event."

temperature. If the latter, then release of greenhouse gas is a direct function of temperature, and one could imagine a positive feedback scenario with sobering implications for the possibility of sudden and dramatic climate change. Either way, the primary controls in effect when global temperature descends into glaciation must be closely allied to whatever is controlling the atmospheric levels of carbon dioxide and methane. The counterpart to this statement, of course, is that global warming is beholden to the same types of controls.

Extraterrestrial Ice Worlds

Icy moons of the solar system provide modern examples of what an extreme glaciation on Earth might have looked like. Probably the best examples are presented by Europa, Enceladus, and Titan. Because of their thin atmospheres and distance from the Sun, these moons experience temperature lows considerably colder than anything that could be experienced on a Snowball Earth. For example, surface temperatures on Titan are about -280°F (-178°C). Nevertheless, these moons provide important potential analogues for past states of our planet.

EUROPA

The Jovian moon Europa is about the size of Earth's Moon and has an active surface that is constantly resurfaced. The smoothest object in the solar system, it is crisscrossed by dark linear features that appear to represent healed fractures in its icy crust and that spread across the surface like veins and arteries. These fractures, along with areas that look like frozen pack ice, indicate that Europa has a thick icy crust supported by a great water reservoir. Ronald Greeley, a scientist on the *Galileo* spacecraft imaging team, notes that there is a lot of evidence for geological activity on the Europan surface. Large plates of the icy crust have broken apart from one another and then refrozen, and it is clear that the pieces "fit together like a jigsaw puzzle." Greeley argues that for this kind of motion to occur, there must be either liquid water or a layer of warm ice beneath the frozen crust that enables the motion. These findings have several implications for understanding the possibility that Earth may have had more extensive ice cover. First, cracks in a global ice sheet would be expected, and would also be expected to heal up if it remained cold. Second, there does not necessarily need to be liquid water beneath the ice sheet to promote fracturing of the sheet; warm ice can serve this function as well. The possibility of warm ice and liquid water on Europa makes it a promising target in the search for extraterrestrial life.

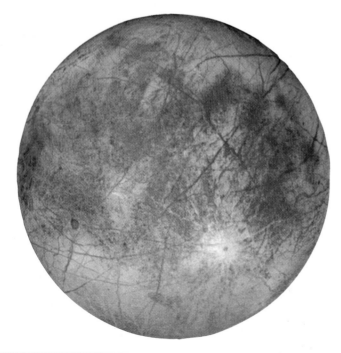

Left: Scientists who subscribe to the Snowball Earth theory often look to the Jovian moon Europa as a potential model. Evidence of geological activity on Europa's surface leads some scientists to theorize that the moon must contain water in liquid form or as warm ice. Top left: Two moons of Saturn, Enceladus and Titan, as seen by Cassini.

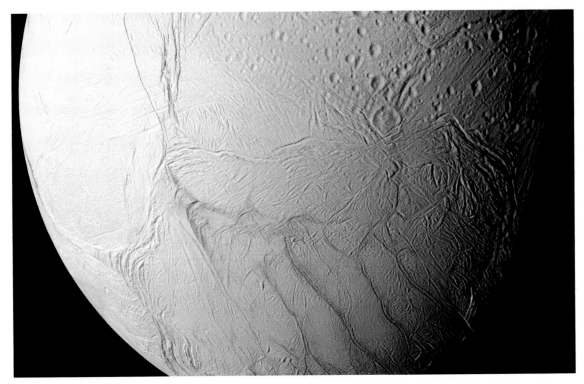

Enceladus, Saturn's icy moon, as seen by the Cassini *spacecraft.*

ENCELADUS

Enceladus, an icy moon of Saturn, is 310 miles (500 km) in diameter. A warm area on the moon's south pole astonished

An artist's conception of Titan's sur-face. Areas once thought to be liquid oceans are now known to be dunes.

planetary scientists when it began releasing a plume of water vapor. Almost like steam coming out of a punctured balloon, the water vapor extends 620 miles (1,000 km) into space. Tidal heating by Enceladus's interaction with Saturn appears to be the heat source that generates the vapor plume, but there is also speculation that the plume is chemically driven by a reaction between oxygen (formed by cosmic ray dissocia-tion of the hydrogen and oxygen in water) and ammonia.

TITAN

Saturn's moon Titan is a huge orange ball some 3,200 miles (5,150 km) in diameter. Isotopic evidence (nitrogen)

indicates that much of Titan's atmosphere has been lost to space. The surface harbors icy bedrock covered with rounded rocks eroded from the icy terrane. Numerous dendritic channels drain into what appears to be the shoreline of a major river channel. On Titan there may be a variation of the Earth's hydrologic cycle, pos-sibly involving liquid methane. An area on Titan once thought to be a lake proved, with higher-resolution images, to consist of hundreds of parallel ridges that are interpreted as dunes. No one knows what particulates compose these dunes; some speculation suggests that the grains forming the dunes are tiny particles of frozen methane.

CHAPTER 9

GEOLOGICAL CATASTROPHES

The study of geological catastrophes and disasters has potentially vast implications for civic planning and disaster preparedness. Many people live under threat of a landslide or debris flow, which can destroy buildings and cut off transportation routes, such as happened in recent years in both Peru and the Philippines. Entire continental plates slowly move on the Earth. At their margins, enormous amounts of energy are released in potentially devastating earthquakes, such as the well-known 1906 San Francisco earthquake. Each of these events has the potential for massive destruction. Underwater landslides and earthquakes cause tsunamis that race across oceans with enough energy to wipe out coastal settlements, as happened in the 2004 Banda Aceh tsunami disaster. Erupting volcanoes fill the atmosphere with debris and cover the land surface in red-hot lava, exposing nearby residents to significant risks. Whether the world will soon be subject to a global warming catastrophe is uncertain. Survey courses of this field, known affectionately in the trade as "shake and bake" courses, introduce students to the tremendous (and sometimes suddenly released) power of various of geological processes.

Left: The haunting beauty of the Lakagigar earthquake area of Iceland belies the fact that it was the site of the second-largest basaltic fissure eruption in historic time. The Lakagigar eruption took place in 1783–84. Top: Devastation in Valdavia, Chile, after the 1960 earthquake that registered a 9.5 magnitude, making it the largest-magnitude earthquake ever recorded. Bottom: Most scientists believe that waste products, released into the air by power stations and factories, play an enormous role in a global-warming trend.

Mass Wasting and Deadly Mudslides

The massive failure of a slope made up of large amounts of soil, rock, and volcanic debris can generate truly destructive mudslides. Heavy rainfall can play a major role in this type of disaster, as can release of volcanic gas in areas upslope from human habitations. If changing climate patterns lead to episodes of heavier precipitation in inhabited areas with soils and other relatively loose materials on slopes such as the sides of canyons, then the frequency of deadly mudslides can be expected to increase.

Above: Edward Hitchcock. Below: A 2005 landslide that hit the track leading to the ancient Incan ruins, led to the evacuation of Machu Picchu in Peru. Top left: Mass wasting occurs when the gravitational force acting on a slope is greater than its shear strength.

MASS WASTING

Mass wasting is the process by which geological materials in solid form move down a slope as eroded blocks, rocks, or sediment grains. This is usually a relatively slow process, as for instance when seasonal frost wedging splits rock fragments away from a bedrock slope. With time, jumbled rocks accumulate at the base of the slope to form what is called a talus slope. Large and dangerous rocks can be moved suddenly downslope in this same fashion, but predicting the timing of such potentially devastating events is beyond our

When Hurricane Mitch struck western Nicaragua in 1998, it triggered massive mudslides, such as this one on the Casita volcano.

current technical ability. For example, the great American geologist Edward Hitchcock (1793–1864) was rather worried about the impressively precarious basalt column bases of Titan's Piazza on the east shore of the Connecticut River in western Massachusetts. But in spite of Hitchcock's concern, this overhanging ledge of rock has not collapsed in the century since Hitchcock's time.

PERUVIAN LANDSLIDES

In October 2005, 1,400 people were evacuated from the ruins of Machu Picchu when a landslide in the night destroyed 1,300 feet (400 m) of railway track leading to the famous archaeological site. This landslide, however, was minor compared to the debris avalanche in 1962 that, traveling at 105 miles per hour (170 km/h), wiped out a large part of the village of Ranrahirca in the Nevados Huascaran region of Peru with a loss of 4,000 to 5,000 lives. This avalanche occurred without warning and without an

obvious seismic trigger. A few years later, in 1970, however, in the same part of Peru (Ancash), a magnitude 7.7 earthquake hit. With a loss of 18,000 lives, Ranrahirca was again heavily damaged, and the town of Yungay was completely destroyed. The 1962 Nevados Huascaran slope failure released 464 million cubic feet (13 million cubic m) of debris, whereas the 1970 event released about 1,765 million cubic feet (50 million cubic m) of material. The avalanche in the latter event sped downhill at an astonishing average velocity of 173 miles per hour (280 km/h), a speed record for a debris avalanche that has not been exceeded to my knowledge. A 1974 rockslide and

debris avalanche destroyed the Peruvian town of Mayunmarca and dammed the Mantaro River. Subsequent failure of the Mantaro landslide dam caused a major flood downstream.

WIND AND WATER

When Hurricane Mitch struck Honduras in 1998, 10,000 people were killed in floods and landslides that were triggered by a combination of 180-mile-per-hour (290 km/h) winds and torrential rains that came down at a rate of 4 inches (10 cm) of precipitation per hour. As the global population has risen, it has become more common for people to live in areas where they are exposed to the risk of debris flows and landslides.

The Philippines mudslide of 2006 left more than 1,000 people dead and completely buried the town of Guinsaugon.

THE PHILIPPINES MUDSLIDE

The Philippines mudslide in Leyte and Cebu in early 2006 left well over 1,000 people dead or missing. The mudslide occurred when a major part of Mount Guinsaugon collapsed, burying the town of Guinsaugon in as much as 32 feet (10 m) of mud. The tragic casualties included a primary school filled with children. Unusually torrential rains probably triggered the mountainside collapse. Witnesses said that it "sounded like mountain exploded, and the whole thing crumbled" and "I felt the earth shake and a strong gust of wind, then I felt mud at my feet."

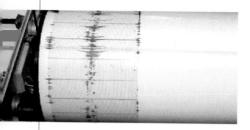

Earthquakes

Tectonic plate boundaries are where the most destructive earthquakes occur (San Francisco in 1906, for

Above: Had the 1906 San Francisco earthquake been measured using today's technology, scientists believe it would have registered a M_W of around 7.8. Below right: A photograph of an early conception of the seismograph. Top left: This seismograph, made by the Kinemetrics Company, was once used by the U.S. Department of the Interior.

example), although destructive earthquakes can be generated anyplace and can strike without warning. The science of earthquake prediction is still in its infancy, but recent advances have been made in determining the frequency intervals for the recurrence of

major earthquakes in particular regions. For example, recent calculations suggest that there is a 62 percent probability for an earthquake of magnitude 6.7 or greater in the San Francisco Bay region during the 30 years beginning in 2003. The 1906 earthquake is often cited as having a Richter magnitude of 8.3, but today large earthquakes are described in terms of their moment magnitude (M_W). The 1906 quake is inferred to have an M_W between 7.7 and 7.9.

EARLY SEISMOGRAPHS

Ancient historical records mention earthquakes as far back as 1800 BCE, and Aristotle provided a classification scheme for earthquakes that placed them into categories depending on the type of movement observed (or felt): side to side, up and down, or a combination of both. Chang Heng in China designed the oldest seismograph in 132 CE; this instrument had the ingenious property of being able to indicate

the direction of the earthquake's first primary impulse or epicenter. The first catalogue of earthquakes was published in 1840, and a few years later, Robert Mallet proposed the establishment of a network of seismic observatories over the surface of the Earth. But it was not until 1892 in Japan, when a compact and relatively easily installed seismograph was designed by John Milne and his colleagues, that it became practical to collect quantitative data on earthquakes on a global scale. Seismographs of this type consist of a vertical steel rod set into a concrete base that extends into the Earth, and an

often-cylindrical heavy mass that is connected to the vertical rod by a suspension wire and a pointed (and hence low-friction) pivot rod projecting from one end of the cylinder. A scribing unit extends from the other end of the massive cylinder, and scratches or writes a needle trace on a rotating recording drum whose base is also fixed into the concrete base.

P- AND S-WAVES

Earthquakes occur when the rocks on either side of a fault are suddenly displaced. This displacement occurs as a release of energy built up along the fault plane because rocks on either side of the fault experience tension as they try to slide past one another. This is called pre-earthquake strain accumulation. When this energy is released as an earthquake shock, a variety of seismic waves are generated. These immediately begin to radiate away from the epicenter. The two primary types of earthquake waves are P-waves and S-waves. P-waves, also called primary waves or compressional waves, are the

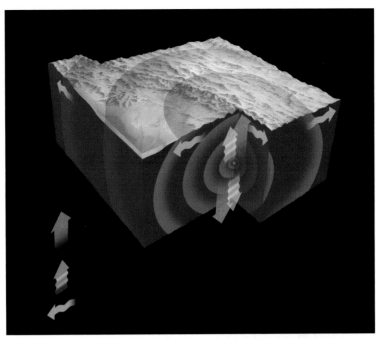

The three types of seismic waves produced by earthquakes: P-waves, or compression waves, are shown in red; S-waves, or shear waves, in yellow; and surface waves, which likely cause the most damage, are shown in purple.

faster of the two and arrive at points distant from the epicenter first. Like sound waves in air, P-waves form concentric bands of compaction and rarefaction around the epicenter. Unlike sound waves in air (because air has no shear strength), but much like a wave bend traveling down the length of a taut rope, S-waves (also called shear waves or secondary waves) arrive after the first arrival of the P-waves. In 1909, Andrija Mohorovicic (1857–1936) used the differing velocities of P- and S- waves to determine that continental crust was much thicker than oceanic.

A contemporary woodcut of the earthquake.

NEW MADRID EVENTS AND THE ANCIENT MIDCONTINENT RIFT

One of the largest earthquakes ever recorded in the contiguous United States was the New Madrid, Missouri, earthquake of February 7, 1812. The main shock was followed by three other major quakes. Two new lakes were formed by the seismic displacement, and the Kentucky Bend was formed by an alteration in the course of the Mississippi River. The New Madrid seismic events are associated with the Reelfoot Rift, a 750-million-year-old extensional fault structure that is associated with the breakup of the Rodinia supercontinent. Unlike other rifts along the margin of North America that became new oceans after the Rodinia breakup, the Reelfoot Rift remained a narrow basin and was subsequently buried by sedimentary cover.

Submarine Landslides

The word "tsunami" is from the Japanese term for harbor (*tsu*) and wave (*nami*). Tsunamogenic submarine landslides (sciorrucks) are capable of generating tsunamis even in the absence of triggering seismic events. The risks of sciorrucks are currently under study as part of a plan to develop an early-warning system; depending on where such events take place, they can lead to truly horrific losses of life. Underwater slides are connected with lateral sedimentation and the formation of what are called "submarine" or "turbidite fans" in deep marine waters.

> **TURBIDITE FLOW SNAPS THE TRANSATLANTIC CABLE**
> The Grand Banks earthquake struck the coast of eastern North America on November 18, 1929. This seismic event triggered a massive turbidite flow that snapped 13 transatlantic cables on its route to deeper water. The snapping of these cables amounted to a gigantic (if unplanned) scientific experiment that showed that the turbidite flow moved at speeds greater than 50 miles per hour (80 km/h).

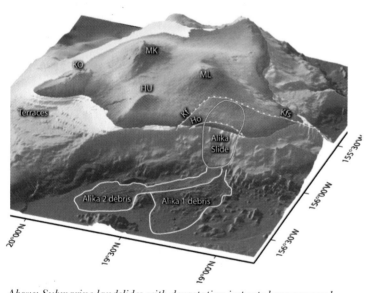

Above: Submarine landslides with devastating impacts have occurred throughout geological time. Shown here is a digital rendering of the Alika 2 landslide, a giant submarine landslide that took place around 120,000 years ago. Some scientists believe that this landslide caused a megatsunami that could explain the marine fossils found onshore in the Hawaiian Islands. Top left: A view of Hilo Bay, Hawaii. Hilo has been devastated by tsunamis twice; once in 1947, and again in 1960.

GREAT SCIORRUCK

Approximately 100,000 years ago, a giant sciorruck set off a tsunami that devastated the nearby islands to an elevation of 1,300 feet (400 m). When this wave reached the eastern shore of Australia, it was still 130 feet (40 m) high. A wave generated by a sciorruck 7,000 years ago struck the northern and eastern coasts of Scotland and devastated lowland areas. This sciorruck is suggested to have been caused by the sudden displacement of more than 960 cubic miles (4,000 cubic km) of marine sediment in a sloping part of the seafloor in the Norwegian Sea. Sciorrucks may be triggered by large seismic events, small seismic events, or may occur spontaneously without any earthquake trigger whatsoever.

LATERAL SEDIMENTATION

Once a sciorruck has been released, the sediments and

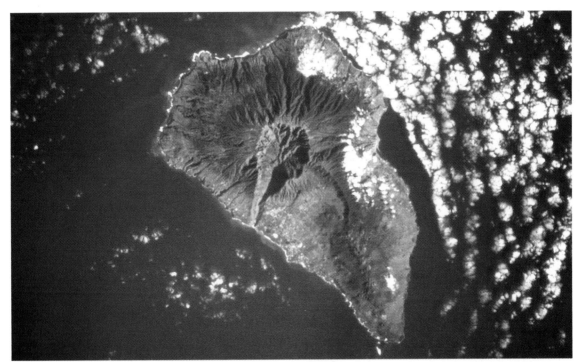

Scientists fear that in the next few thousand years, the volcanic island of La Palma in the Canary Islands in the eastern Atlantic Ocean will erupt. The resulting landslide could trigger a massive tsunami with the potential to devastate the islands of the Caribbean Sea and the eastern coast of the United States.

rocks that compose it move rapidly downslope into deeper water. This sediment slurry, a mixture of saltwater and sediment, is denser than the water that surrounds it, so it stays at the seafloor and can travel great distances, even to abyssal depths. This process is called turbidite flow, and successive events of this nature eventually accumulate to form submarine, or turbidite, fans on the deep seafloor. Sometime such downslope flow, instead of consisting of tiny particles of loose sediment, consists of huge boulders and blocks of lithified or partly lithified sediments from the continental slope or shallower areas on the seafloor.

These blocks can accumulate at the base of a submarine slope to form a kind of deposit known as an olistostrome. Impressive olistostrome deposits, consisting of chaotically arranged blocks of varied origins, are exposed along the California coast at Dana Point. Events of lateral sedimentation, whether involving loose sediment or olistostromal blocks, can be involved in generating sciorrucks. This is because of their lateral-settling as opposed to downward-settling motion.

OUT OF AFRICA?

On the east side of the Atlantic Ocean, off the coast of West Africa, sediments continue to accumulate on a continental shelf that has the potential to generate a truly massive sciorruck. This area should be monitored carefully, for it has potential for generating waves that could race across the Atlantic and wreak destruction on coastal cities and nearshore areas in the Americas. Were an area to be identified off the coast of West Africa that seemed to have potential for generating, or triggering via chain reaction, a gigantic sciorruck, it might be wise to consider steps to create trenches or other devices that could allow the orderly shunting of these coastal sediments to deeper water where they can do no harm.

Seismic Tsunamis

Earthquakes (and volcanic explosions such as the one in 1883 at Krakatau, which lies in the Sunda strait between the islands of Java and Sumatra) in the marine realm can generate huge waves that cross entire oceans at great speeds. Their destructive force is legendary in nations adjacent to the "Pacific Ring of Fire," such as Japan. The aftermath of tsunamis are devastating as well. After the 2004 Indonesian tsunami, more than 40,000 wells in Sri Lankan coastal regions became useless because of contamination with seawater. It will take years for these wells to recover.

HISTORIC TSUNAMIS

The Grand Banks earthquake of 1929 was associated with a tsunami of historic proportions that killed 29 people in a sparsely populated Canadian coastal region. Since then, there have been seven great tsunamis, and of these, four have occurred in the vicinity of the Aleutian Islands. The Aleutian Islands were hit in 1946 and again in 1957, the Kamchatka Peninsula in 1952, and Prince William Sound in 1964. The geological context for these four events is the seismic activity associated with the subduction zone off the south

The 2004 Indian Ocean earthquake and ensuing tsunami have accounted for just under 200,000 deaths, making it one of the deadliest natural disasters in the past 100 years.

THE BANDA ACEH TSUNAMI DISASTER

On December 26, 2004, an enormous wave, triggered by an earthquake in the Sumatra subduction system, devastated the southwestern coast of Sri Lanka, the Andhra Pradesh coast of India, and a long stretch of the Sumatran coast and other areas that were not thought to be at great risk from tsunamis. This great Sumatra-Andaman earthquake added up to a foot (31 cm) of sand on islands in the Maldives.

coast of the Aleutian chain, where the Pacific oceanic crust is sliding into the mantle. The 1946 event destroyed the Scotch Cap lighthouse on Unimak Island, Alaska, and killed nearly a hundred people when the wave reached Hilo, Hawaii. Hilo

was also devastated by the wave from a 1960 earthquake off the coast of Chile. The 1946 tsunami convinced American authorities that an early-warning system was needed, and 1948 saw the establishment of the Seismic Sea Wave Warning

occurred at 5:36 PM local time as a magnitude 9.2 earthquake with an epicenter depth at 14 miles (23 km). A wooden plank was driven through a truck tire by the tsunami at the town of Whittier, Alaska; the wave surge also claimed 122 lives and caused more than $100 million worth of property damage.

THE GREATEST WAVE

The greatest wave in recorded history occurred in 1958 in a fjord called Lituya Bay in southeastern Alaska. This fjord, a glacially carved marine embayment, has steep sides rising 548 feet (1,800 m) above the level of the water. An earthquake struck the area at 10:15 PM, triggering an avalanche on the steep slope, and the debris rushed a kilo-

meter downward into the fjord. This avalanche is technically not a sciorruck because it was initiated above the level of the water. When the landslide material hit the water, it triggered a giant wave that rushed across the fjord and demolished all vegetation to a height of 1,640 feet (500 m) on the other side. The water then sloshed back down, generating a 1,719-foot-high (524 m) wave. Traveling at 93 miles per hour (150 km/h), it overtook a fishing trawler at anchor in the bay. A couple aboard the boat witnessed and, amazingly, survived the catastrophe; the wave snapped the boat's anchor cable, and the vessel, although stood up on its stern, was carried to relative safety as the wave energy dissipated in open water.

Above: The resulting tsunami from the Good Friday earthquake was powerful enough to drive a piece of lumber through a truck tire. Bottom: Taken soon after the 1929 Grand Banks earthquake, this photograph shows houses in Lord's Cove, Newfoundland, demolished by the resulting tsunami. Right: Aerial photographs of the North Shore of Lituya Bay, Alaska, show the extent of the wave's devastation. Prior to the wave, the dense forestation that can be seen at the top of the photograph extended all the way to the shoreline. Opposite page, top left: Tsunami warning sign in Phuket, Thailand.

System. Perhaps the most powerful earthquake in United States history is known as the Good Friday earthquake or Great Alaskan earthquake. It

Channeled Scablands

Ancient flood events are known to have occurred as a result of melting of ice dams and the release of great glacial lakes. The geological evidence in this area led to a great dispute that was resolved in favor of the existence of these ancient floods. During our current episode of global warming, smaller-scale flood events may occur in association with melting glaciers and the lakes that will form as a result behind ice dams or "ice corks."

GEOMORPHOLOGY OF WASHINGTON STATE

Shortly after the end of World War I, in 1918, the American geologist J Harlan Bretz (1882–1981) pondered a highly puzzling series of geological features in eastern Washington State. The region between Spokane and the Columbia and Snake rivers is a terrane consisting of flood basalts with a layer of loess (wind-blown, glacially derived dust). Deep, roughly parallel channels, locally called coulees, are cut into this loess blanket. The entire area has a decidedly odd surface geomorphology, giving rise to its name—the channeled scablands.

SCABLAND FEATURES

A number of anomalous features of the channeled scablands have inspired special geological interest. Basalt outcrops in the coulees are strangely scoured with deep scoring marks or gouges. These scour marks are neither the work of direct glacier contact nor ordinary river erosion. At the base of the coulees are seen discontinuous gravel deposits composed of basaltic and other types of rock transported some distance from their point of origin. Elevated hillocks, consisting of loess, are scattered among the coulees. The coulee channel-ways sometimes cross lofty divides, indicating that an astonishing amount of water must have been available to flow over the divide. Finally, the coulees contain hanging tributary valleys (tributaries that do not cut down to the coulee channel floor), indicating that for at least during the scabland-forming event, the coulees must have been filled with water. Taken

Left: Dry Falls in central Washington State was created when the ice dam that formed the Glacial Lake Missoula gave way, putting areas of the northwestern United States hundreds of feet underwater. Thought to be the most impressive waterfall the Earth has known, Dry Falls was left desiccated when the Columbia River returned to its normal course. Top left: As glaciers retreat, glacial lakes, such as this one in Bulgaria, are formed.

The floodwaters that Bretz believed were responsible for the Columbia Plateau's unique geological features likely occurred as the result of the sudden release of Glacial Lake Missoula. Now dry, it is seen here.

as a whole, these anomalous features point to an unusual mode of formation for the channeled scablands. When aerial photographs of the Columbia Plateau became available, they showed giant sediment ripples, 430 feet (131 m) long and 23 feet (7 m) high, on the floors of several coulees.

NEGLECT OF THE HYPOTHESIS

When Bretz first proposed a catastrophic origin for the floodwaters, his ideas were met with disbelief by the geological establishment. Geologists of his day were simply not prepared to accept such an unlikely seeming catastrophic explanation, especially considering how they had

been weaned on the doctrines of Lyellian uniformitarianism (the principle that natural laws are constant in space and time) and Darwinian gradualism (the principle that change is typically slow, steady, and gradual). And indeed, Bretz's claims seemed extreme. He claimed that 3,000 square miles (7,770 square km) of the Columbia Plateau were swept clean of their silt and loess by the glacial flood, and that 1,000 square miles (2,590 square km) developed gravel deposits derived from flood-eroded and flood-transported basalt. Bretz concluded that it "was a debacle which swept the Columbia Plateau." Challenged by skeptics for lack of a plausible mechanism, Bretz replied that his

interpretation of the scablands "should stand or fall on the scabland phenomena themselves."

VINDICATION

Bretz's analysis gained fresh credence when evidence began to accumulate for a giant, ice-ponded lake called Glacial Lake Missoula. Formed two states away in western Montana, Glacial Lake Missoula is calculated to have released more than 750,000 cubic feet (21,237 cubic m) of water *per second*, a flow volume sufficient to displace 40-foot-diameter (12 m) boulders when the ice dam was breached. Bretz's interpretation is now the preferred explanation for the geomorphology of the channeled scablands.

Volcanes

Volcanic eruptions can be classified into a variety of types, such as Plinian (explosive eruptions that release columns of tephra and gas) and Strombolian (characterized by fountains of basaltic lava). Their destructive power became manifest at Pompeii, with the deadly side-blast of Mount St. Helens, and in the explosion of Mount Unzen in Japan. The remains of these volcanoes will eventually swell with silicic lava again.

VOLCANIC DEVASTATION

Mount St. Helens's explosive eruption of 1980 destroyed a forest of fir trees, and the blast left their volcanic ash–covered trunks all oriented away from the direction of the blast. The Waha`ula Visitor Center in Hawaii Volcanoes National Park was destroyed, literally burned to the ground, by a basaltic lava flow. A deadly lahar (a slurry of water mixed with volcanic debris) formed in 1985 when, during the eruption of the Colombian volcano Nevado de Ruiz, about 10 percent of the volcano's ice cover was melted. The resulting lahars raced down the river channels on the slopes of the volcano and traveled as far as 62 miles (100 km) at the cost of at least 22,000 lives.

BURPING LAKES

Even in a quiescent volcano such as the one that encloses Lake Nyos in Cameroon, emanation of volcanic gases such as carbon dioxide can pose considerable risks. In 1986, gas released by a magma chamber 78 miles (125 km) below the surface of Lake Nyos resulted in a massive cloud of carbon dioxide vapor. Although carbon dioxide is a small component of ordinary air, pure carbon dioxide is denser than air and will flow downhill. The gas cloud from Lake Nyos flowed downward for 15 miles (24 km) with lethal results, leaving 1,700 people dead. The reason for the timing of the formation of the lethal gas cloud is still something of a mystery. The bottom waters of Lake Nyos become naturally charged with carbon dioxide from the gradual input of volcanic carbon dioxide. This natural soda water is denser than the fresh water at the top of the lake and therefore usually stays put at the lake bottom. For some unknown reason, Lake Nyos underwent overturn in August 1986. The bottom waters came to the surface, and in the absence of overlying water pressure, they released carbon dioxide in a fizz of bubbles to form the deadly gas cloud via a 252-foot-high (77 m) geyser at the lake surface.

AIR NAVIGATION HAZARDS

During its final descent into Anchorage in December 1989, KLM flight 867 flew into a volcanic ash cloud just released by the eruption of Redoubt volcano. The ash caused failure in all four of the airplane's jet engines, and the aircraft plummeted downward for eight dreadful minutes while the crew

Above: A limnic eruption, or lake overturn, gives the water of Lake Nyos a cloudy appearance.
Top left: Popocatepetl, an active volcano about 43 miles (70 km) southeast of Mexico City.

The Redoubt volcano in Alaska. In 1989, Redoubt erupted, nearly causing the crash of a passenger aircraft.

frantically scrambled to restart the engines. The encounter with the Redoubt eruption occurred at 23,000- feet altitude, so very fortunately the pilots were able to restart the engines while the plane lost about 15,000 feet of altitude. On landing safely, the plane required $80 million in repairs. Monitoring of active volcanoes, even those incapable of forming an ash cloud as impressive as the one that was released by Redoubt, is important to protect the thousands of passengers and hundreds of millions worth of materiel that are flown daily over volcanoes of the Pacific Rim. For example, Anchorage International Airport was shut down for 20 hours during the 1992 eruptions of Mount Spurr, Alaska.

Oregon's Crater Lake, formed by the collapse of Mount Mazama, is the deepest lake in the world that is completely above sea level.

CRATER LAKE CALDERA AND MOUNT MAZAMA

The Crater Lake caldera in Oregon represents the remains of Mount Mazama, a volcano that underwent catastrophic collapse approximately 7,000 years ago. A caldera (from the Spanish for "cauldron") is a large crater formed by the collapse of a volcano into itself. Islands within the Crater Lake caldera, such as Wizard Island, represent andesitic and rhyodacitic cones and domes that erupted a few hundred years after the caldera collapse.

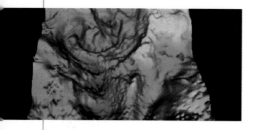

Impact Events

Near-Earth asteroids and comets pose ever-present threats to the planetary surface. The statistical distribution of these events is not only important for future disaster planning but also a useful dating technique for the surfaces of rocky planets and moons. A newly discovered crater in Wilkes Land under the East Antarctic Ice Sheet may help to explain the Permo-Triassic mass extinction event that killed off 95 percent of all marine life 250 million years ago.

TUNGUSKA EVENT

Reminiscent of what happened to the trees on the slopes of Mount St. Helens in 1980, on June 30, 1908, an explosion occurred in central Siberia over the Krasnoyarsk Territory, Irkutsk region, and Yakutiya. Often referred to as the Great Siberian

Above: A colored map detailing the geological structure of the Barringer, or Meteor, Crater in Arizona. This crater was formed when a iron meteorite collided with Earth around 50,000 years ago. Below left: A 1927 photograph shows trees that were felled by the Tunguska event of 1908. Top left: A remote sensing image of the area near Chicxulub, Mexico, where a large asteroid is believed to have struck at the end of the Cretaceous period.

Explosion, it flattened 60 million trees over 830 square miles (2,150 square km). The Tunguska event was not associated with a volcanic eruption, however. The most popular current explanation for this event is that it represents the explosive disintegration of a comet or asteroid about 6 miles (10 km) over central Siberia. Some scientists have disputed this explanation and would rather attribute the craterless blast to an explosion of methane gas released from the Earth or other geophysical event. If indeed caused by a comet or asteroid, the Tunguska event surely shows the devastating potential of impact events. Had the event occurred a few hours later the Earth would have rotated so that the city of Saint Petersburg would have been in the path of the impact.

FUTURE IMPACT

Meteors and other cosmic dust particulates are constantly raining into Earth's atmosphere, giving us the nocturnal phenomenon of shooting stars. The probability that any one of these meteors is large enough to cause damage on Earth is small, but given a sufficient amount of time, large and destructive impact events are a statistical certainty. Indeed, the end-Cretaceous extinction event that killed the dinosaurs is thought to be associated with the impact of a 6-mile-diameter (10 km) asteroid that left the 112-mile-diameter (180 km) Chicxulub Crater in Mexico's Yucatan Peninsula. Cretaceous-Paleocene boundary strata of the Brazos River area in Texas have been a locus of dispute regarding the biotic and sedimentologic influence of this impact event. University of Washington geologist Joanne Bourgeois and her colleagues have argued that they can recognize a massive tsunami deposit at the boundary site on the Brazos. Gerta Keller of Princeton argues that the extinction horizon and the impact event horizon are not the same at the Brazos (in fact, separated by 300,000 years), and her results have produced thought-provoking discussions among geologists, some of whom suggest that the apparent separation could be due to the tsunami deposit.

PLANET PROTECTION

Concerns about disastrous impacts in the past have led to proposals to protect humanity from the next severe impact. NASA has a major project under way to catalogue most of the worrisome objects more than six-tenths of a mile across within the next few years. Related work has led to identification of an asteroid called 99942 Apophis, discovered in 2005 and calculated to make its closest approach to Earth on April 13, 2029. This 1,000-foot (307 m) object has only about a 1 in 45,000 chance of actually hitting Earth in 2036, but these are worrisome-enough odds to claim the attention of asteroid-hazard experts. One possibility might be to tag 99942 Apophis with a transmitting beacon to track its motion with greater accuracy. Were it to prove a significantly increased risk for impact, a defense mission could be

A rendering of the Don Quixote mission, a system for deflecting asteroids on target to collide with Earth that is currently being studied by the European Space Agency.

mounted to deflect the course of the object and save Earth a serious collision. The longer we wait, however, the harder it will be to do this, and if we wait until 2029 it may well be too late to deflect the path of Apophis.

The circular basin at the top of this image is the Moon's Mare Orientale.

MARE ORIENTALE

Mare Orientale (the "eastern sea") is not actually a sea, but instead an 810-mile (1,300 km) multiring impact structure. From Earth the dark side of the Moon mostly obscures it, but in 1962 Gerard Kuiper and William K. Hartmann at the Lunar and Planetary Lab, University of Arizona, projected their best photographs of the Moon onto a three-foot-diameter (1 m) white globe to form a rectified image. When this was done, the ghostly edge of the giant impact structure became visible for the first time.

CHAPTER 10

A HISTORY OF GEOLOGICAL THOUGHT

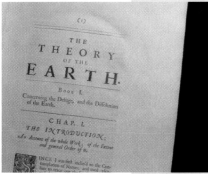

Left: Considered by many to be the founder of modern geology, Charles Lyell was an early exponent of scientific gradualism. Top: Popular geologic thought posits that asteroid impacts with Earth are "regular" events in the context of geologic time. Bottom: A chapter from Thomas Burnet's The Sacred Theory of the Earth, *an early, yet scientifically unsubstantiated, attempt to understand Earth's structure.*

Geology was born alongside the concept of a young world that experienced a great flood, reached adolescence with the concept of actualism ("the present is the key to the past"), and is reaching maturity with a newfound appreciation of catastrophic change. Early views of change on the Earth's surface often focused on catastrophic events. As geology emerged as a modern science, the focus shifted to understanding the operation of incremental changes over time, leading to a distinctly gradualistic bias that held sway over the field for more than a century. Geologists are now prepared to appreciate both gradual and catastrophic processes, and our understanding of geology and evolution are colored by the concept of punctuated equilibrium. In other words, things stay pretty much the same (equilibrium) until major change comes along (punctuation). It has led to better understanding of change in the natural world through time, for it involves a very reasonable interplay between processes that occur quickly and those that move at a more stately pace. It is also a question of perspective. A giant asteroid impact might seem to be a rare, unexpected event, but viewed over a long stretch of geologic time such events must be expected.

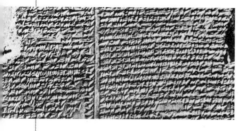

Mosaic Chronology and Catastrophism

The earliest visions of Earth's history were decidedly catastrophic, or rather cyclic, involving periods of sudden destruction followed by intervals of great fertility.

RENEWAL BY DISASTER

After a 1766 earthquake wiped out most of the native inhabitants of Cumana (the first city founded in America by the Spanish, in what is now Venezuela), Alexander von Humboldt (1769–1859) reported that the Indians reacted "following the concepts of ancient superstition, with festivals and dancing to celebrate the destruction of the world and the approaching epoch of its regeneration." Thus a period of extreme fertility was to follow the time of destruction.

FLOOD TRADITIONS

The Chinese have a concept of a general deluge that goes back to 4000 BCE This tradition, dated to the earliest period of Chinese history, seems to be associated with a major disruption to Chinese agriculture.

The account of the Noachian deluge of the biblical tradition can be traced back to Sumerian mythic cycles. The possibility that this is connected to an actual geological disaster has been a topic of lively speculation. Hans Baldung Grein's 1516 painting, *Die Sintflut* (The Flood), is used as the cover illustration of *Environmental Disaster and the Archaeology of Human Response*, a 2000 book edited by University

Alexander von Humboldt

of New Mexico professors Garth Bawdena and Richard Martin Reycraft. The painting accurately conveys the terror that accompanies such geological disasters. In his *Sacred Theory of the Earth* (1681), theologian Thomas Burnet (1635–1715) provided an explanation for the Great Flood. First, the Earth's crust develops a gigantic crack. Second, water rushes up from the giant reservoir in the abyss

Left: Flood stories are common to many cultures. According to Hindu lore, the god Vishnu comes to Earth in fish form to warn Manu, the King of Dravida, of an impending flood that will destroy all life on Earth. Like Noah, Manu is given the task of preserving the "seeds of life" and is spared from death. Top left: An ancient Assyrian tablet, found in Nineveh, describes a mythical flood, much like the one encountered by Noah in the Bible.

Burnet theorized that the "Great Flood" occurred when the Earth's crust cracked and water from its core rose to push apart the continents. He points to major mountain chains, such as the Andes, as the remnants of the fissured crust.

to inundate the Earth. Finally, the waters retreat, leaving the continents as we see them today.

Burnet interpreted the major mountain chains such as the Andes as the raised edges of the sundered crust. For Burnet, the Earth was smooth in its paradisiacal condition after emerging from an initial state of chaos. The Great Flood crack led to complete flooding of the Earth's surface, followed by draining of the waters and formation of continents and mountain ranges as noted above. In Burnet's scheme, the Earth will in the future undergo a conflagration, be returned to its smooth idyllic state, and finally be transfigured to a luminous body that shines like a star, thus completing the sequence of changes from "dark chaos to bright star."

Although the Bible never actually states how old the Earth is, biblical or Mosaic chronology has been used to provide an estimate of its age. The calculated value was much too young, but at least it was a start that could be improved upon, as indeed proved to be the case. By John Phillips's time in the 1860s, it had become clear to many that the Earth was not just thousands but millions and indeed hundreds of millions of years old, and the question was just how far back this deep time stretched.

Neptunism and Volcanism

The charismatic early geologist Abraham Gottlob Werner (1749–1817) provided the science with both the first global stratigraphy and a theory for the origin of the stacked sequences of rock that he was able to recognize in different parts of Europe. Werner was evidently loath to publish, but the main statement of his theories appeared in print in the 1780s. In his stratigraphic classification scheme, he divided the rocks of the world into four layered units; he later added a fifth unit to account for rocks that evidently belonged between original stratigraphic units one and two.

WERNER'S STRATIGRAPHY

Werner's stratigraphy began with primitive rocks (*Urgebirge*), consisting mostly of felsic igneous rocks such as granites, plus some metamorphic rocks such as schists. Next in the stack were the transitional rocks (*Ubergangsgebirge*), sequences of primarily marine limestones and clastic rocks plus mafic igneous layers that would now be attributed to the Late

Above: Abraham Gottlob Werner. Top left: James Hutton's engraving of the granitic rock at Glen Tilt where he first came to believe that granite was cooled magma that had risen from below the surface of the Earth.

Paleozoic. Next was a series of twelve recognizable layers that were grouped together as the secondary or stratified series (*Flotzgebirge*). These strata included both terrestrial and marine deposits, plus mafic igneous rocks and coals, that would now be assigned to the Mesozoic and Cenozoic. Above

the *Flotzgebirge* is the alluvial group (*Aufgeschwemmte Gebirge*) that consist of relatively recent, weakly lithified coarse and fine clastic sedimentary rocks ranging from conglomerate to crumbly clays. And finally, topping all, were volcanics (*Vulkanische Gesteine*), principally lava flows and tephras that, at least locally, covered everything else.

NEPTUNISTS VERSUS VOLCANISTS

Werner's stratigraphy allowed recognition of the general relative age of any particular body of rock, and was all-inclusive, namely, no rock types were left out. It had a disadvantage, however, and that was that Werner had taken the side of the neptunists versus the volcanists in the debate over the origin of basalt. Neptunists, such as the French geologist Jean Étienne Guettard (1715–86), argued that basalt was not an igneous rock but rather was precipitated from a liquid such as seawater at low temperature. Volcanists, such as pioneer field volcanologist Nicholas Desmarest (1725–1815), insisted that basalt has formed from an originally very hot volcanic magma. Werner was impressed by the layered character of many

European basalt occurrences. For a neptunist, this meant that the rock formed as a layer of precipitate crystals; for a vulcanist, it meant that the magma that eventually formed the basalt had flowed some distance to form a layer of what is now called, somewhat ironically, flood basalt.

SETTLING THE CONTROVERSY

The controversy continued even after the Scottish geologist James Hutton (1726–97) made critical observations at Glen Tilt in the Grampian Highlands, and the island of Arran and in Galloway. At Glen Tilt, he found dikes of red-colored (and hence potassium-feldspar-rich) granite intruding into and cross-cutting dark mica-rich schists and other rock types. He took this as compelling evidence that the granite has risen from below as a hot magma and been injected into the cool, overlying rock. Thus, Hutton is known as a plutonist, and today large bodies of granite formed at depth are called plutons.

The controversy over the origin of basalt lingered for some time,

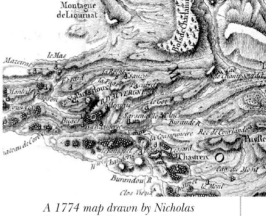

A 1774 map drawn by Nicholas Desmarest indicating various columnar basalt formations near a volcanic hill, or puy.

and, in 1799, another Scottish geologist, Richard Kirwan (1733–1812), claimed to have proved the neptunist origins of basalt by finding marine shells in basalt exposed at Portrush in Ireland. Close inspection of the site, however, showed that the fossils occurred in a shale that had been metamorphosed to a dense, hard rock by being adjacent to an injected layer of basalt. The basalt layer itself was chilled (that is, had smaller crystal sizes along its edges) by being in contact with the comparatively cool fossiliferous clastic sedimentary rock.

Basalt columns found at the Giant's Causeway on the northeast coast of Ireland in County Antrim. For years after its official "discovery" in 1693, it had been disputed whether the Causeway had been created by nature, by man, or by a mythical giant. In 1771, Desmarest identified the basalt columns as having been formed by volcanic activity.

Steno and Uniformitarianism

Many commentators mark the advent of modern geology by the publication of Danish geologist and paleontologist (and later Roman Catholic bishop) Nicolaus Steno's 1669 book best known by the first word of its Latin title, *Prodromus*.

STENO'S PRINCIPLES

In his study of Tuscan geological history, Steno articulated three principles of stratigraphy that are the foundations of historical Earth science to this day. The first is original horizontality, the idea that rocks are laid down originally in more or less flat-lying layers. The second is original lateral continuity, the concept that the original layers were laterally extensive in spite

of now being cut up by river erosion, faulting, and other natural processes. The third is called superposition, the idea that in a sequence of layered rocks the oldest layer is at the bottom of the stack. The third principle is easy to comprehend, in fact seems obvious, and layering processes occurring and observable today must have had counterparts in the past. This leads to the doctrine of actualism.

A NEW ERA

With the 1830 publication of the now-classic *Principles of Geology* by British geologist Charles Lyell (1797–1875), the science of geology entered a phase that emphasized the gradual over the catastrophic

and the uniform over the irregular. Thus was codified the geological doctrine of uniformitarianism, a viewpoint promoted

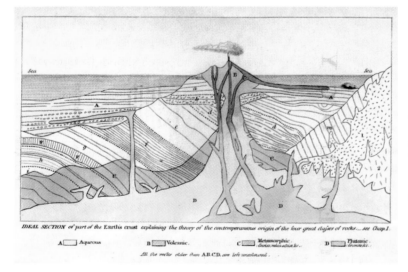

IDEAL SECTION of part of the Earth's crust explaining the theory of the contemporaneous origin of the four great classes of rocks... see Chap.I

A ☐ Aqueous   B ☐ Volcanic.   C ☐ Metamorphic Cortex.mica-schist &c.   D ☐ Plutonic Granite &c.

All the rocks older than A.B.C.D. are left uncoloured.

Left: The frontispiece from the second edition of Charles Lyell's Principles of Geology. *In this book, Lyell stressed uniformitarianism over catastrophism and argued that understanding the geological processes of the present was the key to understanding the geological processes of the past. Above: A page from Nicolaus Steno's* Prodromus, *a book many consider to be the birth of modern geology. Top left: Nicolaus Steno's ideas of original horizontality, lateral continuity, and superposition can easily be applied by geologists studying exposed rock formations such as this one.*

According to methodological uniformitarianism, geological processes at work today, such as the creation of ripples along this shoreline, have been at work in similar fashion during every stage of the Earth's existence.

early on by James Hutton when he envisioned the Earth as having "no vestige of a beginning, no prospect of an end." This doctrine of uniformitarianism still plays a role in modern geology. It may be divided into two aspects, substantive uniformitarianism and methodological uniformitarianism.

SUBSTANTIVE UNIFORMITARIANISM

Substantive uniformitarianism is a fallacy, for it denies the Earth a history and in its extreme form requires the cycles of mountain uplift and erosion to be extended back into time ad infinitum. Hutton, for instance, viewed the Earth in this way, as is evident in the comment cited above. Lyell went so far with substantive uniformitarianism as to argue that, as soon as appropriate environmental conditions return to the Earth's surface, dinosaurs would again roam the land and pterodactyls would again take to the air. Lyell also felt at one time that continued research into the fossil record would show that human remains go back to the times of the dinosaurs or even earlier! Charles Darwin did not make this mistake, but he was strongly influenced by Lyell's uniformitarianism. He therefore had a decided theoretical preference for gradual evolutionary change among organisms through time.

METHODOLOGICAL UNIFORMITARIANISM

Methodological uniformitarianism, on the other hand, is the idea that geological processes that can be observed on Earth today operated in a similar fashion at any time in the Earth's past. For example, sand ripples that form today in a streambed formed in the same way in the Cretaceous, the Carboniferous, the Cambrian, or the Archean. This leads to the geological doctrine of actualism, the concept that "the present is the key to the past." Actualism has universal support in the geological community, and in fact is required if we wish to attempt to make accurate inferences about ancient environmental conditions by observing rocks formed during the times that those conditions held sway. This is not to say that there cannot have been, by present standards, unusual conditions on the Earth's surface in the remote past. Such conditions must be studied from a nonactualistic perspective, and form some of the most challenging problems in the geological sciences today because there are no modern analogues with which they may be directly compared.

Continents Adrift

For as long as relatively accurate world maps have been available, youngsters and others with open minds have noted the fit between the coastlines of South America and Africa. Burnet, of course, had a giant crack in the Earth's crust to release the waters of the Great Flood, and in 1858, French geographer Antonio Snider-Pellegrini added pressure to the Burnet scenario, publishing a book in which he argued that abyssal waters beneath the Earth forced their way upward through an enormous north-south fissure that then became the North and South Atlantic.

CLOSING THE ATLANTIC

But serious scientific consideration of the problem had to await the attention of German meteorologist Alfred Wegener (1880–1930). Convinced of the inadequacy of contemporary theories for the configuration of the Earth's crust, he mustered paleontological data and other evidence such as the fit of the coastlines to make the case that the continents were once joined. Concerning the opposite sides of the Atlantic, Wegener claimed that it "is the same as if we put back together the pieces of a torn newspaper, matching the edges and checking to see that the lines of print run continuously across." His evidence included glacial strata

Above: A drawing by Antonio Snider-Pellegrini depicts the opening of the Atlantic Ocean. He believed that continental drift was caused by a north-south rift in the supercontinent, which caused abyssal waters to flow up through the fissure, creating the Atlantic. Top left: A German postage stamp commemorates the scientific achievements of Alfred Wegener.

and fossil and living plants and animals on the now separate southern continents. These forms, Wegener argued, could not have crossed the Atlantic in its present extent. For example, Wegener's *Mesosaurus* looks like a miniature crocodilian less than a few feet long, and would not be capable of surviving a swim (or even a float) from Africa to South America. Wegener also found evidence for living animals (particularly distinctive types of Southern Hemisphere earthworms and freshwater fish) that could not have crossed the Atlantic, either. Tracing the trajectories back into time of all these lines of evidence, he united all continents into a single supercontinent he called Pangea. Wegener found an ally in the work of South African geologist Alex du Toit (1878–1948), who argued that the mass of continents was split into northern (Laurasia) and southern (Gondwana) halves by an equatorial seaway he named Tethys.

EARTH-MOVING IMPLICATIONS

Wegener's ideas were generally met with disdain by his geological contemporaries. The reasons for this were primarily twofold. First, geology as a science was still beholden to the uniformitarian assumptions inherited from the work of Charles Lyell in the 1830s. Continental drift seemed to be an affront to

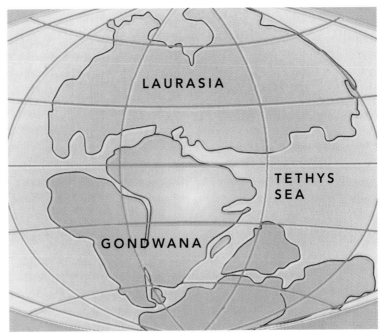

In Alex du Toit's conception of Pangea, a seaway he labeled Tethys separated the enormous landmass into two halves, Laurasia and Gondwana.

the notion of a fixed and regular Earth surface. Certainly mountains were uplifted and eroded down again, but continents in motion (even slow motion) was a bit much to ask. Second, Wegener had no plausible mechanism for drifting the continents. His critics rightly pointed out that trying to plow, say, Africa through the solid mafic rocks of the seafloor crust would break the continent to

bits. Wegener did not live long enough to solve this geophysical problem, for he died on a 1930 expedition to Greenland whose purpose was to measure its rate of motion away from Europe. A hint of a solution was already available in the 1920s, however, with the ideas of geoscientist Arthur Holmes (1890–1965) about convection currents rising in the Earth's mantle beneath the continents.

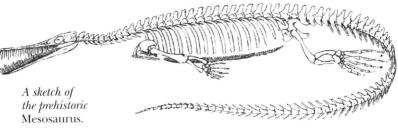

A sketch of the prehistoric Mesosaurus.

Catastrophism, Old and New

The doctrine of catastrophism, out of favor at least since the 1830s' publication of Lyell's *Principles of Geology*, came roaring back in the 1980s with new data supporting the impact-extinction hypothesis for the demise of the dinosaurs.

A REVIVED DOCTRINE

Catastrophism had, of course, never completely vanished from the realm of geological investigation. The Chamberlin-Moulton planetesimal theory had a decidedly catastrophic aspect, and particularly the single-impact origin hypothesized for the Moon must have represented the most violent disaster in the entire history of the planet. The 1980s also saw development of the concept of "impact frustration" of the origin of life: the idea that life could not possibly gain a foothold on this planet until the era of truly great planetesimal impacts had ended.

Crater counts on the Moon, in conjunction with radiometric dating of the Apollo mission lunar samples, showed that the era of truly great impacts had ended about 3.8 billion years ago. After this time, according to the theory, the surface of the early Earth would cool enough to first, allow permanent oceans to form, and, second, to allow life's beginnings in whatever environments hosted its origin, be they hydrothermal vents or someplace else.

THE ALVAREZES

In 1979, physicist Luis Alvarez (1911–88) and his son, geologist Walter Alvarez (b. 1940), began circulating a manuscript that reported the discovery of a geochemical spike in the iridium content of a thin layer of sedimentary rock called the boundary clay. The boundary clay marked the Cretaceous-Tertiary boundary in a sequence of marine limestones exposed in Gubbio, Italy. The boundary in this stratigraphic section was precisely defined on paleontological grounds using microfossils of foraminifera, several species of which served as index fossils. The Alvarezes and their coworkers found that the content of iridium in the boundary clay was elevated above background

Above: An image of Crater 302 on the surface of the Moon. By studying samples from Moon craters, scientist have been able to determine that the era of planetesimal impacts ended 3.8 billion years ago, allowing the Earth's surface to cool enough to begin supporting life. Top left: Scientists believe that after the Earth's surface was sufficiently cooled, life may have begun to form around hydrothermal vents, such as this one.

Above: Luis Alvarez worked with his son Walter Alvarez to develop the asteroid-impact theory. Alvarez proposed that the high concentration of iridium at the K-T boundary line indicated the impact of an extraterrestrial body. Below: A rendering of the Chicxulub crater as it might have looked not long after impact. The remains of the asteroid, mainly in the form of iridium, can be found around the world in a layer of rocks that mark the boundary between the Cretaceous and Tertiary periods.

levels by at least 60 times. They attributed this huge geochemical marker to a post-impact dust cloud formed of the remains of a giant bolide (asteroid or comet) that slammed into the Earth 65 million years ago. Iridium is rare at the Earth's surface (most of Earth's iridium is with the iron that flowed to the core), and the high content in the boundary clay suggests an extraterrestrial source.

CONFIRMING DATA

This announcement triggered a spirited debate in the geological community, but as always, geologists went to the rocks for more data. The iridium spike also turned up in nonmarine strata of the Raton Basin, New Mexico, dramatically bolstering the image of the impact theory. Quartz grains with shock fracturing that

had been seen previously only at nuclear test sites started turning up in the boundary clays. And finally, geologists using an airborne gravimeter discovered a huge, 186-mile-diameter (300 km) impact crater buried under marine sediments just north of the Yucatán Peninsula. Further investigation turned up evidence, in archived oil-crew core samples, of a thick breccia formed of broken rock and frozen globs of mafic glass. These were indicative of an impact event large enough to penetrate down to the basalts and gabbros of the Cretaceous seafloor crust under the Gulf of Mexico. With these findings, belief in a major impact event as a causative agent for the end-Cretaceous mass extinctions became widespread (but by no means universal). Catastrophism had returned.

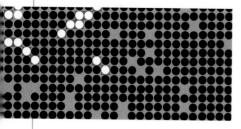

Vernadsky and Gaia

A striking thing about the history of life is that, for something like three and a half billion years, the Earth has neither become too hot nor too cold to support life. This is so in spite of all the catastrophes that have assaulted Earth's surface over this time, from asteroid impacts to dramatic and sudden climate change. This fact alone so impressed scientists James Lovelock (b. 1919) and Lynn Margulis (b. 1938) that they became proponents of what is now called Gaia theory. Gaia theory states essentially that life and its living processes have acted, over the course of geologic time, to maintain the surface of our planet in a habitable state. This is so, according to Gaia proponents, because a variety of feedback mechanisms exist between life and its environment, and these feedback mechanisms act to prevent planetary surface conditions from departing from those conditions conducive to the long-term continuation of life on this planet.

Above: Vladimir Vernadsky. Below: James Lovelock poses in front of a statue of Gaia. Top left: A screen shot from James Lovelock's computer program Daisy World. This program was designed to illustrate how biotic processes can regulate the Earth's climate.

GAIA'S ANTECEDENTS

The Gaia concept has an interesting relationship to the ideas of the late Russian geochemist Vladimir Vernadsky (1863–1945). Known for his research into the geochemistry of silicates and other types of minerals, Vernadsky became extremely impressed with the influence of living processes on the chemistry of rocks on the Earth's crust. Vernadsky felt that life was involved in virtually every type of geological process

Lynn Margulis, an early proponent of Gaia theory.

on Earth, and he went so far as to express the idea that life completely controls the geochemistry of surface rocks—in other words, life makes geology. For Vernadsky, living processes are highly opportunistic, as likely to throw the global climate system out of balance as to stabilize it and to help it maintain some type of equilibrium. Vernadsky spoke about this almost explosively opportunistic character of life as the pressure of life. For Vernadsky, life was so effective in influencing and modifying its environmental surround, through its "media forming" abilities, that Earth must really be viewed as a living planet. The geology on Earth would not be the same without the microbes and other organisms of the crust that enhance weathering rates, bore into rocks and minerals, release acids, and so on.

COMPLEMENTARY VIEWS

So here we have two complementary views of the role of life on Earth. The Vernadskian view sees life as an aggressively active series of catalytic processes that can change the atmosphere (oxygen crisis, two billion years ago), create new rock types (coal that forms from Hypersea), and even tear down mountains (biotic enhancement of weathering). The Gaian perspective sees a gentler system where, by their very growth and reproduction, organisms help to maintain a happy climate on the Earth's surface.

According to Gaia theory, all organisms are responsible for maintaining a stable and life-sustaining environment on the Earth's surface.

DAISY WORLD

Lovelock created a special computer program called Daisy World to promote the Gaian view. In Daisy World, there are two types of daisies, one white and one black. Black daisies absorb sunlight and warm climate. White daisies reflect sunlight and cool climate. If the simulation world gets too hot, the black daisies die off (heat prostration) and the white daisies prosper. If it gets too cold, the black daisies expand at the expense of the white daisies. Thus the temperature of Daisy World is kept at a constant over a fairly wide range of solar luminosities. The Vernadskian and Gaian views are complementary rather than being in conflict. Vernadsky emphasized life's propensity to expand rapidly (a type of positive feedback); Gaia theory emphasizes the stabilizing nature of controlling feedback (negative feedback) in biotic processes. Both types of feedback are known to occur in nature.

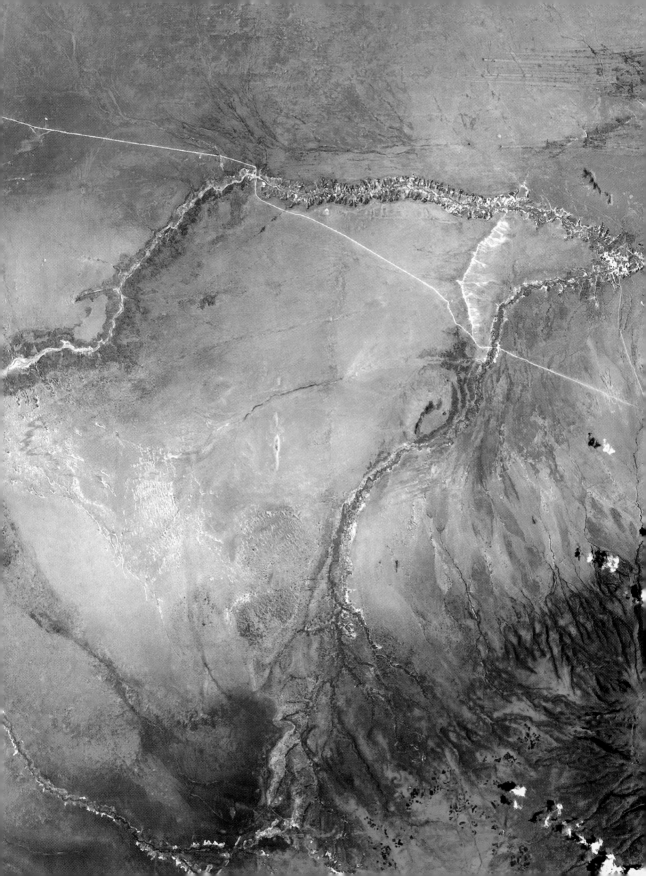

CHAPTER 11

THE PLATE TECTONIC REVOLUTION

Left: The Great Rift Valley is home to one of the most important archaeological finds of all time. This satellite image shows Olduvai Gorge, where Louis and Mary Leakey discovered fossilized remains of both Paranthropus *and* Homo habilis. *Top: The study of plate tectonic theory has greatly improved our understanding of volcanic activity, including its role in the evolution of atolls. Bottom: In Iceland, a bridge connects the Eurasian and North American continental plates. In between is the Alfagja Rift Valley.*

At first, critics were skeptical about the mechanisms of continental drift, but the great success of plate tectonic theory has led to advances in all areas of geology, from understanding the evolution of atolls to prospecting for natural resources. From convection rolls in the mantle to magma chambers beneath mid-ocean ridges, many of the general processes driving plate tectonics are understood. Nevertheless, many questions about mechanism remain. For example, what role, if any, do interactions between the Earth's core and mantle play in influencing the rate of continental drift? What causes supercontinents to break apart? Do carbonate rocks on subducting slabs play a role in "lubricating" the subduction process? Another rapidly developing area of research involves plate tectonic's relationship to both major alterations of Earth's climate and to major advances in the complexity of marine and terrestrial biotas. Can breakup of a supercontinent trigger a major glaciation? Can supercontinental breakup induce the evolution of new types of animals and plants? What role if any does the biosphere play in maintaining the tectonic system? All of these questions can and will be addressed by earth scientists who study continental drift and plate tectonics.

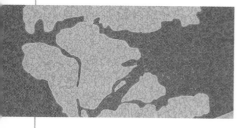

Reconstructing Rodinia

Piecing together ancient supercontinental jigsaw puzzles is a great geological enterprise that helps us to understand the history of Earth. Fossil distributions played a critical role in the early attempts to reconstruct ancient supercontinents, and the technique still finds use in this role today.

A PROTEROZOIC SUPERCONTINENT

The first supercontinent to be recognized and named was the one formed of the present-day southern continents and now called Gondwana. Eduard Suess (1831–1914) originally called the supercontinent Gondwanaland, but since *Gondwana* is a regional term from India meaning "land," Gondwana without the "-land" is the better term for the supercontinent. In the late 1960s, earth scientists began to suspect the existence of an even earlier supercontinent. James W. Valentine and Eldridge M. Moores published a paper in 1970 with the earliest known attempt to reconstruct this continent. They showed a simple, circular supercontinent with a single large rift valley running through its center. This later broke up to form satellite continents, each one of these

Above: Eduard Suess. Left: Putting the continents into their earlier "super" forms presents a jigsawlike puzzle to geologists. Top left: The present-day continents were a single landmass during the Triassic.

in the Valentine and Moores cartoon looking like wedges cut from a pie.

FINDING THE PIECES

It has proved to be a considerable task to put all the pieces of this supercontinent back together. For example, using paleomagnetic evidence, J. D. A. Piper published supercontinental reconstructions in the 1980s that used most of the continental blocks. His reconstruction, however, showed considerable paleogeographic separation between Australia and northern

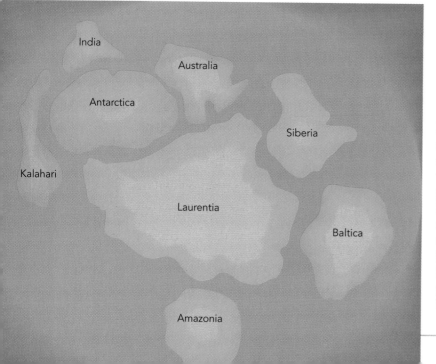

India

Australia

Antarctica

Siberia

Kalahari

Laurentia

Baltica

Amazonia

A Glossopteris fossil found in Australia. The wide distribution of Glossopteris fossils in the Southern Hemisphere is important evidence for the existence of the Gondwana supercontinent.

RECONSTRUCTING RODINIA AND GONDWANA

Paleobiogeographical evidence in the form of distribution of fossil organisms and the fit of the continents are the two primary criteria for reconstructing supercontinents. Gondwana was reconstructed based on the distribution of the Permian reptile *Mesosaurus*, the plant *Glossopteris*, and distribution of ancient glacial deposits. Animals and plants had not appeared in the fossil record when Rodinia formed, so its reconstruction is more reliant on geological evidence such as the fit of continental margins and similarities in strata sequences of appropriate age. The fossils *Dickinsonia*, *Tribrachidium*, and *Spriggina*, however, are best known from Australia and Baltica. These two areas, widely separated today, must have been closer together during the time of Rodinia, and this paleontological evidence has helped the reconstruction efforts.

in the Proterozoic. Siberia has also been placed near the western margin of North America, and pieces of Africa and South America seem to account for the continental fragments that rifted away from the present-day-coordinate eastern and southern North American margin. Mark A. S. and Dianna L. Schulte McMenamin named this supercontinent Rodinia in 1990. Radiometric dating demonstrated that Rodinia formed about a billion years ago and began to break apart around 750 million years ago. As continents moved away from North America at the core, what is called a passive continental margin formed around the perimeter of North America. This passive continental margin, a habitat of warm shallow seas and gradual seafloor slopes, became host to a wide variety of marine organisms and appears to have been associated with the evolution of the first large types of animals.

GONDWANA FORMS

After the Rodinia breakup, North America was left in a splendid isolation that did not end until it collided with Europe and Africa hundreds of millions of years later. Plate motion and collision after the breakup allowed Antarctica, Australia, and the disparate pieces of Africa and South America, plus a few other pieces such as India, to come together as the supercontinent Gondwana. Gondwana later became the southern half of Pangea.

Europe (Baltica), and this was in conflict with paleontological evidence that showed (and still shows) a strong paleobiogeographic link between the Proterozoic fossils of these two continents. The gap between Australia and Baltica can be closed by invoking a tectonic model, first proposed by Charles W. Jefferson in 1978, that placed Antarctica and Australia together as the missing piece off the west (present-day coordinates) coast of North America. Jefferson

and R. T. Bell published a map in 1985 that showed Australia and North America juxtaposed. Jefferson even predicted that Ediacaran fossils would be found in the Backbone Ranges in northwestern Canada, and they were in fact later discovered there.

AT THE CENTER

Jefferson and Bell's map has subsequently been confirmed, and the task since then has been to identify the other continents that surrounded North America

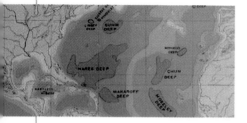

Tectonic Essentials

Subduction zones may occur on the edges of continents or out at sea, and where they occur, old oceanic crust is being consumed and melted. In order for the Earth to maintain a constant radius, new ocean crust must be generated, and this occurs at mid-ocean ridges. Therefore, the Earth's crust undergoes a continuing cycle of destruction and renewal.

TETHYS

Alex du Toit did maintain a disagreement with Alfred Wegener over the supercontinent Pangea. Du Toit felt that two supercontinents were involved, Suess's Gondwana in the south and what du Toit called Laurasia in the north, consisting of North America, Greenland, and most of Asia/Siberia. We now know that both men were right. Pangea did exist as a single supercontinent,

and was split in two to form northern and southern halves, Laurasia and Gondwana respectively. The split occurred along a pivot point that was close to the Iberian Peninsula, and opened up a great east-west equatorial seaway called Tethys. Tethys

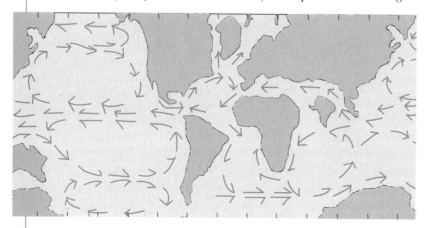

Left: Around 45 million years ago, the Earth was girdled by an equatorial seaway, named Tethys by Alex du Toit. Tethys allowed for the distribution of similar marine species throughout the world. Above: The Meteor, *the German research vessel that confirmed the existence of a mid-ocean ridge. Top left: A view of the mid-Atlantic ridge. Opposite: This image was created by scientists aboard the* Meteor *and is believed to be the first three-dimensional model of the mid-Atlantic ridge.*

allowed warm-water marine organisms of similar species to distribute themselves from East Asia to West Africa, from East Asia to Mexico, and was associated with upwelling and nutrient-rich equatorial seas that generated great hydrocarbon deposits (in the form of oil) in the Middle East. In order for Tethys to form, and for the great continental masses of Gondwana and Laurasia to move away from each other, new oceanic crust had to be formed.

THE MID-OCEAN RIDGE

In an expedition sent out to study ocean chemistry, the German ship *Meteor* made an astonishing discovery as it crossed the Atlantic on a number of voyages between 1925 and 1927. Based on the ship's echo soundings en route, taken every 5 to 20 miles (8 to 30 km), a bathymetric anomaly began to emerge. There evidently was a broad, roughly linear ridge running north-south down the middle of the Atlantic Ocean. This confirmation of the presence of a mid-ocean ridge eventually became a key element in the development of plate tectonic theory. The ridge represents the zone of creation of new oceanic crust, and its elevation above the abyssal seafloor is a direct result of the fact that the basalts and gabbros that form the ridge are still relatively fresh, hot, and therefore buoyant with respect to surrounding rocks of the oceanic crust. When the rate of new seafloor formation is high, the ocean ridges tend to be particularly hot and buoyant with respect to the Earth's mantle. This displaces vast quantities of seawater; so global rise in sea level (transgression) is associated with fast continental drift.

SEAFLOOR RESURFACING

Around 1900 it was felt that ocean floor research was going to be a great boon for understanding the deep history of Earth, because the ocean basins were thought to hold a more-or-less complete stratigraphic record of the entire planetary history. This hopeful expectation met with grief when, much later, it became apparent that the sediments on the modern seafloor are all Jurassic or younger. Since new ocean crust forms at the mid-ocean ridges, old crust must be destroyed to make way for the new. This occurs at subduction zones, where old and cold seafloor crust dives into the mantle and is melted and destroyed in a grand process of planetary recycling.

The Wilson Cycle

Due to the closing of some ocean basins from subduction, continents once apart are drawn into continent-to-continent collisions. These collisions lead to the world's great orogenies, or mountain-building events, and the entire process is associated with seafloor spreading.

SEAFLOOR SPREADING

In an ambitious paper of 1962 that he termed a work of "geo-poetry," geologist Harry Hess (1906–69) outlined his seafloor-spreading hypothesis where new ocean crust is formed at the mid-ocean ridges and old ocean crust is destroyed at subduction zones. The confirmation of this hypothesis came just the next year in simultaneous insights by Canadian marine geologist Lawrence Morley and British geophysicist Fred Vine, who realized that the recently mapped pattern of parallel magnetic stripes (each stripe with its own characteristic width) on the seafloor represented zones of magma erupted during the same phase of Earth's magnetic polarity. For long stretches of geologic time, Earth has alternated between two polarity states—normal (with the north magnetic pole near Earth's north rotational axis) and reversed (with the magnetic pole near the geographic South Pole). For reasons that are not yet fully understood, but

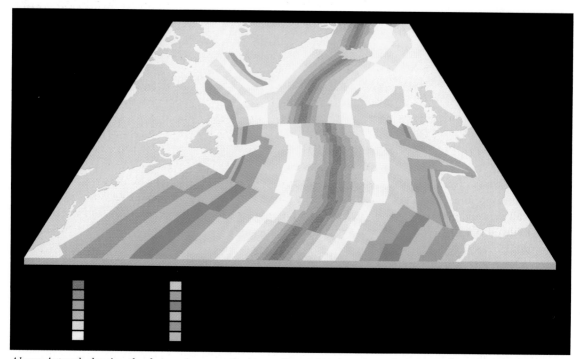

Above: Artwork showing the changes in magnetism in seafloor volcanic rocks as the Earth's magnetic field changes over time. Seafloor spreading, combined with plate tectonics, helps to explain continental drift, or how the continents move relative to one another. Top left: The Himalayan mountain range as seen from the International Space Station.

probably involving rotational dynamics of Earth's core that generates the magnetic field, Earth's magnetic field periodically reverses its polarity. The seafloor magnetic stripes in the oceanic crust are parallel to the mid-ocean ridge. Contemplation of this fact may have led Morley and Vine to the critical observation that the pattern of stripes on one side of the mid-ocean ridge is a mirror image of the stripe pattern on the other side. This single observation, indicating as it does that the edges of oceanic plates are added to at the ocean ridges by successive accretions of new magmatically derived crust, perhaps did more than any other observation to convince geologists of the truth of continental drift and plate tectonics.

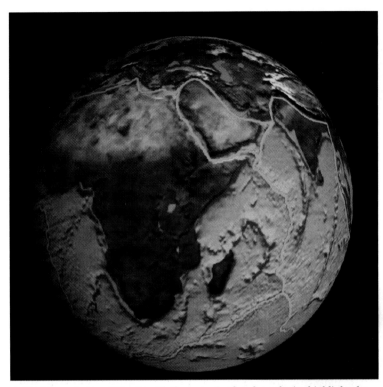

A NASA rendering of Earth with the tectonic plate boundaries highlighted.

PLATE TECTONICS

The theory of plate tectonics was born when the Canadian geologist J. Tuzo Wilson (1908–93) realized in 1965 that large-scale motions in Earth's crust were concentrated along continent margins, mountain chains, and mid-ocean ridges. These features delineated what we now call tectonic plates. Wilson realized that these semirigid plates jostled one another along their edges, and in doing so were responsible for generating most of the surface geology of the planet. Turning to the paleontological record, he noted that the distinctive Middle Cambrian trilobite genus *Paradoxides* was common to Europe and to a thin stretch of territory along the eastern coast of the United States, but was absent from Middle Cambrian rocks of the North American interior. Wilson noted, correctly, that the terrane on the east coast of the United States (now referred to by many geologists as Avalonia) had a distinctly old-world signature, and he went from there to make the inference that Avalonia was once on the other side of an ocean that was the predecessor to the Atlantic. The main implication of this finding was that the Atlantic had not simply opened once as a result of continental drift (as everyone since Antonio Snider-Pellegrini had argued), but that the modern Atlantic Ocean was the end product of a series of ocean openings and closings.

THE HIMALAYAS

The Himalayan mountain chain and the Tibetan Plateau are the result of a continent-to-continent collision between India and Asia. This collision is associated with some of the fastest rates of continental drift ever measured, which partly explains why the tallest mountains today are found in the Himalayas. A large part of the Tethyan seaway was destroyed by this continent-to-continent collision.

Island and Volcanic Arcs

Volcanic arcs form as the result of the upward flow of magma generated by melting of the subducted slab in a subduction zone. When occurring out at sea (by subduction of an oceanic plate beneath an oceanic plate), they are referred to as island arcs.

SUPERCONTINENTAL CYCLE

The insights from J. Tuzo Wilson led to elucidation of what we today call the Wilson cycle. The Wilson cycle describes the cyclic opening and closing of ocean basins as supercontinents such as Pangea break up and the resulting smaller continents are subsequently reunited, albeit never in exactly the same configuration. As a supercon-

tinent breaks up, it does so along faults in the continental crust that lead to the formation of rift valleys. And when dispersed continental fragments rejoin, they do so by means of subduction zones that consume the oceanic crust that separates the tectonic plates with continental crust. The supercontinental cycle is driven by thermal patterns and by density contrasts between the continental crust, the oceanic crust, and the Earth's mantle.

Supercontinental breakup is apparently driven by heat buildup beneath the continental crust, which acts as a sort of insulating blanket that traps heat flux from the Earth's mantle and core. The supercontinent is torn asunder by rift valleys, some of which will open up via seafloor spreading into new ocean basins with hot, buoyant basaltic ocean crust. Two continents with an older ocean between them, for example, will start to converge when the old ocean crust,

Left: A satellite image of the Hawaiian Islands. These islands were formed by the action of basaltic shield volcanoes. Above: Because of their liquid, basalt-rich magma, shield volcanoes, such as the Kilauea volcano in Hawaii, are not as steep as composite volcanoes. Top left: The Kluchevskoj volcano in Russia is an example of a composite volcano.

having grown cold and more dense, begins to slide into the mantle along a subduction zone. This recycled ocean crust will be remelted and erupted as a volcanic arc.

VOLCANIC ARCS

As a cool oceanic slab descends into the mantle, its relatively high water content (it was in intimate contact with seawater, after all) encourages rapid melting of the volcanic rocks and associated seafloor sediments, as soon as the descending slab becomes sufficiently hot. The magma that is generated as a result moves upward and is eventually erupted as part of a steep-sided composite volcano. Composite volcanoes have relatively steep sides because of the high silica content of their constituent magmas, in distinct contrast to basaltic volcanoes, such as the Hawaiian Islands, which have very gradual slopes (shield volcanoes) resulting from their very fluid, basaltic magmas. The volcanoes erupted from the magma generated by the melting subducted plate form a chain paralleling the subduction trench that delineates the descent of the slab. These trenches are the deepest parts of the ocean, yet they are at the same time the cause of lofty volcanoes not far distant.

If the volcanoes form

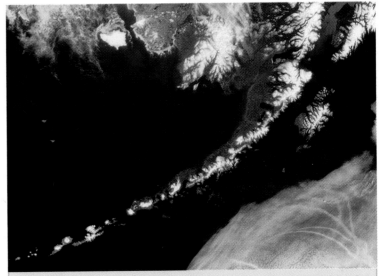

The Aleutian Islands, seen from the Terra satellite. The volcanic island cluster gets its curved shape from the descent of the Pacific plate.

THE ALEUTIAN ISLAND ARC: WHY IS IT CURVED?

The Pacific plate dives into the mantle along the Aleutian trench, which takes the form of a long curve because of the Earth's roughly spherical shape. As the Pacific plate descends, it generates a curved form in much the same way as a knife cutting into an apple. The result is the curved trench and island arc at the Aleutians.

through continental crust, then a volcanic arc such as the Andes Mountains is formed. If the volcanoes form at sea, as for instance when an ocean plate (old and dense) subducts beneath another ocean plate (generally younger, warmer, and less dense), an island chain of volcanoes, called an island arc, is formed. The Aleutian Islands are a classic example of island arc formation.

The Andes Mountains were formed by tectonic forces at the subduction zone where the Antarctic and Nazca plates have slid beneath the South American plates. This ongoing geological event still triggers earthquakes and volcanic eruptions.

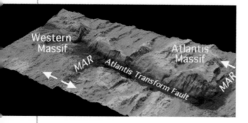

Faults, Rifts, and Ridges

Where sections of the Earth's crust meet, the surface is crossed by faults, rifts, and ridges.

TECTONIC FAULTING

Tectonic motions may be generally classified into two categories: tensional motion and compressional motion. Tensional motion occurs when tectonic plates pull away from each other. They are associated with the kind of faults (called normal faults) typically associated with the formation of rift valleys. These tensional forces will tend to stretch and thin the crust. Compressional motion occurs when tectonic plates move toward each other. Since the two converging plates cannot occupy the same place at the same time, compressional tectonic forces are associated with a lot of buckling and crumpling of rock bodies, which is expressed as folding and thrust faulting. Thrust faulting occurs when one sheet of rock is thrust above or on top of another type of rock. A subduction zone, in fact, can be considered a gigantic type of thrust fault. An inclined zone of earthquake epicenters called the Benioff Zone outlines the downward motion of the subducting slab.

OBDUCTION AND OPHIOLITE TERRANES

The concept that a denser sheet or slab of rock will subduct beneath a less dense sheet of

Above left: A rendering of the Great Rift. The rift runs 3,100 miles (5,000 km), from Syria to Mozambique. Left: Ophiolites, such as this one found in Gros Morne National Park in Newfoundland, Canada, form when a continental plate subducts underneath a denser oceanic plate. Top left: A computer image of a transform fault along the mid-Atlantic ridge. Transform faults help the mid-ocean ridges adhere to the spherical shape of the Earth.

This clearly defined strike-slip fault in the Italian Alps marks the boundary of a transform fault.

terranes are of great geological interest, because a number of them are pre-Jurassic in age and thus represent samples of oceanic crust that would otherwise be completely lost to science because of the crustal recycling process associated with subduction.

TRANSFORM FAULTS

In addition to discovering the supercontinental cycle that bears his name, J. Tuzo Wilson also realized that, by necessity, faulting was going to be associated with mid-ocean ridges. Once again, Earth has a roughly spherical surface, and therefore there must be a way to "curve" the mid-ocean ridges that would otherwise be straight on a flat planet. This bending of the ridges, perhaps best seen in a map of the entire Great Rift, is accomplished by relatively short faults called transform faults. These transform faults offset short segments of mid-ocean ridges and permit the ridges to follow the curved central axis of an ocean basin. The term *transform fault* is applied to larger scale features as well. Wilson used the term *transform* to describe any place where one plate slid past another. Thus, a transform plate boundary is neither convergent nor divergent, but is marked by a roughly vertically oriented fault called a strike-slip fault.

rock serves as a general rule for plate tectonics. A dense oceanic plate will subduct beneath a less dense ocean plate, and a continental plate, being less dense than either, will ordinarily be unable to be subducted beneath an oceanic plate. Exceptions to this general rule can occur, however, and in relatively rare circumstances a slice of oceanic crust (usually broken off from an ocean slab that is subducting in the usual manner) will ride up over an area of continental crust. The deposit thus formed is called an ophiolite. An ophiolite is a sequence of basaltic igneous and associated sedimentary rocks brought to the surface by the obduction process. Ophiolite

A view of the San Andreas Fault on central California's Carrizo Plain .

THE SAN ANDREAS FAULT

The San Andreas Fault is a deadly yet fascinating case of a mid-ocean ridge that, in a sense, transformed into a giant strike-slip fault. This famous fault along the coast of California is the answer to the following question: "What happens when a subduction zone tries to swallow a mid-ocean ridge?" The solution to the problem is a giant strike-slip fault, formed because the mid-ocean ridge (being composed of hot, fresh, buoyant basalt) was unable to go down into the trench. The subduction zone was thus put out of business, and the Pacific plate had no option but to slide against the North American plate along the northwest-southeast trending San Andreas Fault.

Plate Tectonics on the Seafloor

Many marine environments are directly influenced by their plate tectonic settings. Among the best known yet least visited of these are the deep-sea trenches such as the Mariana (or Marianas) Trench, the deepest point in the ocean at 35,798 feet (10,911 m). These trenches are in the places where a subduction slab of oceanic lithosphere dives into the mantle. Trenches are usually associated with a shallow sea on the other side of the resulting volcanic arc, called a back-arc basin.

CONTINENTAL BORDERLANDS

Stretching and thinning of the continental crust is not only associated with supercontinental breakup. Even the margins of continents with predominantly convergent tectonics can develop what are called horsts and grabens, in other words, parallel rift valleys (grabens) with elongate ridges (horsts) between that separate the valleys. When conjugate horst-graben sets occur right at a continental margin, as they do off the coast of Southern California, the grabens will become fairly deep marine basins, such as the Santa Barbara basin, and the horst may stick up as islands such as Santa Catalina Island. This type of a geological and geographic province is referred to as a continental borderland. Marine basins in a continental borderland can accumulate highly organic-rich sediments for several reasons. First, if the basin is separated from the mainland by one or more other basins,

Above: Santa Catalina Island is an example of horst sticking up from a deep marine basin. Top left: A small island in the Hawaiian chain.

then only the very finest-grained clastic sediment is going to reach the basin, because coarser sediment will be intercepted by the basins closer to the shore. Second, borderland basins tend to be silled, which means that the bottom of the basin is like a bowl or bathtub, and ocean currents can access this part of the basin only by flowing down over the basin rim, or sill. Strange symbiotic marine organisms inhabit the regions beneath the sill. This depositional environment, with its abundance of organic-rich sediments, is an ideal setting for the accumulation of sedimentary rocks that can act as the source rocks for petroleum and natural gas resources.

The Puu Puai cinder cone in Hawaii Volcanoes National Park. The island of Hawaii is the youngest of the chain that emerged from a basaltic lava outflow.

HOT SPOTS AND SEAMOUNTS

At various times and at various points in the Earth's mantle, magma will well up and erupt at the same spot for tens of millions of years. These spots remain fixed with regard to the Earth's interior and are not displaced by tectonic plate motions in the crust above them. Like a lava lamp that gets turned on and off periodically, hot spots generate eruptive phases that can be responsible for generating massive amounts of igneous rock. And like an upside-down sewing machine needle punching though thick fabric, hot spots will leave separate, periodically erupted piles of volcanic debris and lava as the plate moves past. The Hawaiian Island chain formed in just this fashion, each island in the chain emerging as the massive outflow of basaltic lava finally reached the sea surface. The youngest island to emerge in the chain is the big island of Hawaii. It is no accident that Hawaii Volcanoes National Park is on the southeastern side of the island, for this is the part of the island that is currently closest to the position of the hot spot. The Hawaiian hot spot is even now in the process of creating what may eventually become a new island, but so far has only been built up to the level of a seamount. This is the Lo´ihi Seamount, located to the southeast of the big island.

CHAPTER 12

GEOLOGY IN THE FIELD

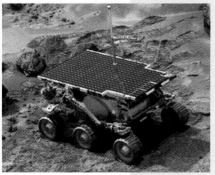

Left: The classic image of fieldwork is shown in The Geologist, *an 1860 oil painting by Carl Spitzweg. Top: The geologist's most basic tool—the rock hammer. Bottom: The geologist's tools can also be high-tech. For example, the rock abrasion tool on NASA's Mars Exploration Rover* Spirit *performs soil-brushing experiments right on the Martian surface.*

Using a variety of tools from rock hammers to GPS to Mars rovers, geologists collect and analyze rocks, bore cores through mud, and create geological maps. Field investigations are fundamental geological activity; in an important sense, all geological research begins in the field. Field investigations are as varied as on-foot rock-collecting trips, aeromagnetic or gravimeter studies from aircraft, submersibles diving in the open ocean, and examination of rocks exposed in mine shafts hundreds of feet below the surface. A certain amount of risk is associated with research in the wilderness; a field geologist may be chased by poisonous snakes, surrounded by javelinas (wild piglike animals), or find his or her tent destroyed and food supplies ransacked by an Alaskan bear. Yet, a reasonable amount of prudence in the field can reduce the risks. Of equal importance is making sure that the requisite permission from the appropriate landowners to access a site of interest is in hand. This can be crucial for work in foreign countries, and the prudent field geologist will be sure to have the blessing of local governmental authorities before beginning work. With this said, it is true that most people have no objection to geologists conducting field research nearby.

Strike and Dip

Measurements of the geometrical attitudes of layers of strata are essential for studying the structural geology of any particular region. Strata may be faulted and folded into anticlines (arch-shaped folds in rock in which rock layers are upwardly convex) and synclines (arch-shaped folds in rock in

and most volcanic deposits (such as lava flows and volcanic ash falls known as tephra) are formed as horizontal sheets of material. After the lava has cooled or the sediment has been lithified, local Earth motions can lead to a wide variety of departures from this originally horizontal condition. Perhaps the most common

STRIKE

When this tilting occurs, the stratal layer in question develops what is called an attitude. The attitude consists of two components: the strike and the dip. The strike is an imaginary line that represents the intersection of the plane of the tilted bed with the horizontal plane—that is, a horizontal plane tangential to the surface of the Earth at the point of investigation (since we know that the Earth is spherical, not flat). Taking the tilted table analogy again, imagine that the room containing the tilted table was filled with enough water to partly immerse the tabletop. The strike line on the tabletop will be the intersection of the water level and the surface of the tabletop. Geological strike is essentially the same thing: the direction of the intersection of the plane of rock with the horizontal plane. Strata surfaces are rarely perfectly planar, of course, because of natural irregularities on the surface of a

Above: In principle, geographic layers are laid down in a horizontal fashion. A number of geological processes can disrupt this natural pattern, however, as this tilted limestone strata demonstrates. Below: The Brunton compass can be used to measure the strike of a bed. Top left: Pahoehoe lava's ropy surface and bizarre shape are due to the fact that it is formed by a series of lava lobes continually breaking out from a layer of cooled crust.

which rock layers are downwardly convex), and can even be overturned. In accordance with the stratigraphic principle of original horizontality, most strata

alteration of the originally horizontal state is simple tilting. In this situation, the stratal layer tilts as would a table if you cut down two legs from one side.

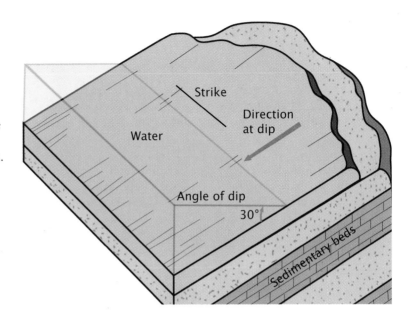

layer of sediment, lithified or not. Geologists can nevertheless estimate the strike of beds with remarkable precision, that is, to within a couple of degrees of the true value. This is done using a transit compass called a Brunton. To measure strike, the geologist can sight along the strike of a bed by looking into the edge of the stratum in question. This is like sighting along the edge of a broken pane of glass until you are looking parallel to the flat surfaces. The direction you are looking defines the strike,

and this is then measured by the compass. Another method involves setting a clipboard or other flat surface against the rock as an extension of its upper surface (bed top) or lower surface (bedding plane sole). The Brunton or other type of compass with a spirit level is then placed against the carefully positioned clipboard surface, and the strike is read directly as the compass is oriented against the board to a horizontal position.

DIP

The dip is easily measured as the angle that the stratal bed makes with the horizontal plane. Dip is always perpendicular to strike and is measured in the downhill direction on the tilted slab. Were a drop of water to be spilled on the clipboard in the second method described above, the water would flow downward and its track would define the direction of dip.

A geologist points out the distinctive shape of an anticline.

ANTICLINES AND SYNCLINES

In addition to being tilted, rocks may be folded. When a planar body of rock is folded, its attitude varies continuously over its surface. In a strongly folded rock sequence, where the rocks are bowed upward, like a wrinkle in a carpet, they form what is called an anticline. If the rocks bow downward, they form a syncline. Anticlines and synclines can occur as conjugate sets in rumpled regions of the Earth's crust (such as the Zagros Mountains in Iran); the result is called a synclinorium.

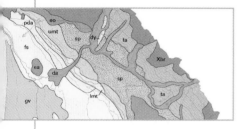

Geological Maps and Cross-sections

Beautiful graphical productions allow geologists to see into the Earth and to read its geological history. Stratigraphic columns allow for precise correlation of stratigraphic sequences in different regions. Of particular importance is how these columns allow recognition of gaps in the sequence called unconformities.

MAPPING THE EARTH

In order to create a geological map, geologists are required to define what are called mappable units. These are particular layers of strata or other types of geological bodies (such as igneous intrusions) that are easily recognizable based on their color, texture, mineral or fossil content, or other features. Once

the map units are defined, it then becomes possible to begin drawing the map. The most important symbols on a geological map are solid, thin lines that are called contact lines. These lines are drawn to show where one type of rock is juxtaposed against another. In a typical case, a tilted, tabular rock unit such as a layer of sandstone will have an upper and a lower contact shown on the map. For ease of viewing, geological maps are typically colored, with each rock unit bearing a distinctive hue. Our sandstone unit mentioned above might then be shown in yellow on the map. Within this yellow zone of the map, representing the outcrop pattern of the sandstone in question, will be plotted the strike and

dip symbol for the sandstone at points where it can be measured in the field. In other words, wherever the geologist can measure a strike and dip, he or she will plot a T-shaped symbol that consists of a long line (the strike line) and a shorter perpendicular line (that shows the direction of dip) at the precise point on the geological map where the measurement was taken. A number associated with this T symbol indicates the dip angle in degrees.

STRATIGRAPHIC COLUMNS

Stratigraphic columns are graphical constructions used to denote the sequence of strata in any particular region. They are most commonly applied to sedimentary rocks, but they can be used to describe other types of layered rock sequences, such as stacked lava flows. In a stratigraphic column, the rocks are shown in proper stratigraphic order, with the

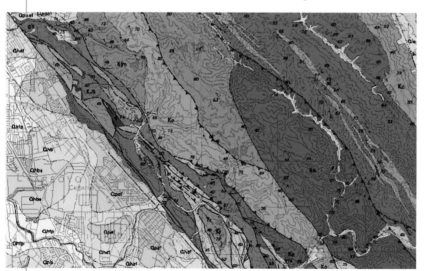

Left: The steepness of the dip or tilt of a bed is represented by the number found next to the strike line. The steeper the tilt of the bed, the higher the number on the map. Top left: In this close-up of a geologic map, different rock types are represented with different colors. The colors represent the immediate bedrock that is exposed in a specific area.

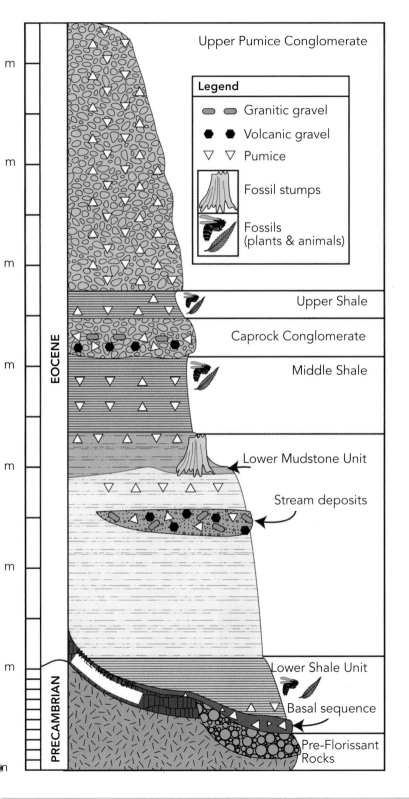

oldest at the bottom (unless the rocks have been overturned by folding or faulting). Subsequent strata are shown on top of the bottom layer, and layers are added in succession until the top of the sequence is reached. The thickness of each layer in the column represents its relative thickness to the other units in the sequence. No tilting or folding is shown; the rock sequence appears in its presumed originally horizontal position. In some stratigraphic columns, more-resistant layers are shown projecting farther to the right to indicate their greater resistance to erosion. Often in "strat" columns, as they are called, pattern symbols will be used to indicate the various rock types that occur in the section. These can include a stipple pattern for sandstone, a block pattern for limestone, a short dash pattern to show shale, and small V symbols to show volcanic layers. The data for stratigraphic columns may be measured from the rocks directly, using a yardstick device called a Jacob's staff, or columns may be taken indirectly from a geological map using an analytical geometry technique that yields an estimated thickness for each rock layer. If there is a break in sedimentation in a sequence, this is shown as a sinusoidal line in the section known as an unconformity.

Stratigraphic columns are like geologic timelines. According to the theory of original horizontality, the rock types on the bottom layers are older than those on the top.

Geospatial Technologies

New computer tools are changing the way that geologists go about making their maps. Image processing of satellite images provides information that is difficult or impossible to acquire on the ground. New techniques in data processing and other fields are dramatically enhancing the speed and precision with which geological field investigations can take place, and, for obvious reasons, are essential for the study of planetary geology.

GEOGRAPHIC INFORMATION SYSTEMS

Geographic Information Systems, or GIS, refers to a host of techniques and technologies used to manage and analyze digitally referenced geographic information.

GIS includes the geographic data itself, along with the hardware and software needed to analyze the data. The technology used to capture the data is part of GIS as well. GIS is used to create a digital mockup of the world called a geographic database, or geodatabase. GIS also has the ability to output this geodatabase in a process known as geovisualization, where maps that include additional information, such as pie charts, population density, or location of geological features, are used to access the information in the complete geodatabase, which is likely to be too complicated to be rendered on a simple two-dimensional map.

GIS is also able to upload, recombine, and analyze already existing datasets—an ability called geoprocessing.

LAYERS OF DATA

Advanced analytical tools can be used to manipulate data from existing datasets, overlay particular geographic features, and to output desired datasets, particularly customized maps and associated data.

As an example, consider a study intended to assess the dangers of potential pesticide use on permeable soil. A digital map of the study area watershed will be linked or clipped to the geology of the

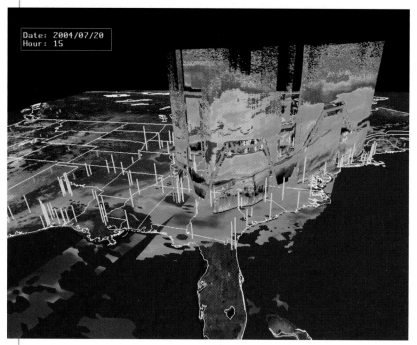

Date: 2004/07/20
Hour: 15

Above: A GIS analysis of the southeastern region of the United States shows the complex layering of data that is now available through computer imaging systems. Top left: Satellite imaging has helped scientists create more accurate geologic maps by introducing new data that would otherwise have gone unnoticed.

area to form a geologic map. To this, another layer will be added showing permeable soil types. In a separate operation, all land uses in the region of interest will be linked to potential pesticide use in the area. This dataset will then be combined with the watershed/geology/soils data to render the geodatabase, which can then be used as a powerful tool for the intended study.

GEOLOGICAL APPLICATIONS

We are beginning to see very fruitful fusions of approaches from traditional geology (for example, tectonic models) and geoprocessing. For example, remotely sensed images using visible, near-infrared, and mid-infrared channels are being used to bridge the gap from regional to local scales in the inference of geological structure. In a 2005 article published in the journal *Remote Sensing of Environment*, P. C. Fernandes da Silva, J. C. Cripps, and S. M. Wise used a combination of remote sensing techniques and empirical tectonic models to infer geological structures that would have otherwise been unknown. Such research strategies promise a great, and greatly useful, series of information and data interchanges between earth sciences and the environmental sciences. Geoprocessing is an inherently interdisciplinary tool because it is based on the utilization and recombination of existing datasets.

GIS illustration showing the terrain elevation levels near Tokyo, Japan.

This image of Cardiff, Wales, taken with a near-infrared radiometer, is part of European Space Agency's research on land cover and vegetation.

THE SPECIAL UTILITY OF INFRARED RADIATION

Remotely sensed images use separate parts of the electromagnetic spectrum, including visible light, near-infrared, and mid-infrared or mid-IR. The spectral channels in the infrared are in many instances providing information in remotely sensed images that is otherwise invisible.

Sample Collection and Preparation

In addition to being an inherently field-based science, geology is also highly dependent on proper collection and management of specimens and samples. From sections of ice cores taken from a fast-melting glacier to dinosaur bones encased in plaster for safe transport to slender rock cores taken from another planet by a sample-return space mission, great care must be taken to collect, prepare, and curate geological specimens. Paleontologists are, of course, well aware of these requirements, for in some cases the most informative fossils also tend to be the rarest and or the most fragile. A single delicate specimen can in many instances mean the difference between a correct geological interpretation and an interpretation that is either incomplete, missing essential details, or just plain wrong.

LOCALITY DATA

Only slightly less important than the specimen itself is an accurate record of the exact site where any particular specimen was taken. This location, known as the specimen's provenance, needs to be recorded with as much care and precision as is necessary for the geological research project under way. These requirements can vary from project to project. For example, in a pioneering study of the stratigraphy of a new region, it may be sufficient to record the formation and location of an isolated piece of rock that contains, say, a fossil that helps to determine the age of the rock. The exact position of the fossil within the formation under study will be difficult to determine at first if the fossil and its matrix are collected from float, that is, small rocks eroded from the main outcrop body by processes of mass wasting. It is better to collect fossil specimens from outcrop (placed within a well-defined stratigraphic sequence) than from float, but in areas where fossils are rare, float specimens may be the only fossils that are readily available.

PALEOMAGNETIC SAMPLING

In a study of the paleomagnetism of ancient rocks, however, much more detailed locality

Above: A paleontologist painstakingly unearths fossilized bones and documents their exact location at a dig in France. Top left: An ice core sample being retrieved from a drill at the Byrd Polar Research Center.

information is required. A rock sample (or drilled rock plug) taken from, say, a basalt flow for paleomagnetic study of its remnant magnetization or residual magnetic signature, must be very carefully labeled as to its spatial orientation in three dimensions before it is removed from the rock outcrop. Float specimens of basalt, unless it was clear exactly where the specimen broke off from its parent outcrop, would obviously be nearly useless for paleomagnetic study, as would poorly oriented samples taken from an outcrop.

SPECIMENS FOR SALE

The question of whether rocks, fossils, and minerals should ever be for sale has implications for the preservation of important specimens. Some purists might argue that these treasures from the Earth are the property of all, our scientific legacy in a sense, and should never be bought and sold. In practice, however, from gem and mineral shows to eBay Internet auctions, such items are bought and sold in massive quantities. A related question is what role amateur collectors should play in the excavation of potentially important geological materials. This question is, in a sense, a double-edged sword. On one edge are unsophisticated collectors who, like grave robbers pillaging a burial site, damage important localities and remove specimens of historical significance. On the other edge of the sword are highly responsible amateur collectors who carefully record their locality and other data, and, without hesitation direct important new finds to professional scientists. Paleontology is particularly dependent on the efforts of such collectors, and the Paleontological Society each year

Above: Responsible collectors are careful to record the locality data of their finds in order preserve the scientific value of their specimens. Below: Gems such as this amethyst quartz are valued for their use in jewelry and attract destructive gem poachers.

gives an award to collectors who have been particularly helpful to the profession. It is safe to say that at a minimum, amateur collectors should carefully record the locality data for all specimens that they collect in order to protect the specimens' potential scientific value. Also, it is important to obey all applicable laws. For example, it is a federal crime to remove vertebrate fossils from public lands in the United States.

GLOSSARY

A

ABYSSAL PLAIN Any low-relief area of the seafloor with a depth greater than 6,500 feet (2,000 m).

ACHONDRITE A stony meteorite with a composition resembling that of Earth basalt.

ACTUALISM The principle that present-day processes can be used to properly interpret events of the geological past.

ALBEDO Reflectivity of a planetary surface or outer atmospheric surface. High-albedo surfaces reflect much solar energy back into space.

ALKALI FELDSPARS A group of feldspars characterized by potassium and sodium content.

ALLUVIAL FAN A low-relief cone-shaped deposit of gravel and sand formed by a stream where it passes out of a canyon or narrow valley.

AMMONITE Any member of an extinct group of coiled cephalopods; most abundant during the Mesozoic.

AMPHIBOLE A common group of rock-forming minerals characterized by double chains of silica tetrahedra and dual prismatic cleavage at angles of 124 degrees and 56 degrees, respectively.

ANDESITE An igneous rock intermediate in composition (particularly in terms of silica content) between a basalt and a rhyolite.

APATITE An important, green or brownish phosphate mineral with formula $Ca_5(PO_4)_3(F, Cl, OH)$.

ARAGONITE A carbonate mineral with formula $CACO_3$. Distinguished by its needlelike crystals that are hexagonal in cross-section.

ARCHEOCYATH Any member of a group of Cambrian spongelike creatures characterized by a porous calcareous skeleton.

B

BACK-ARC BASIN A linear basin that forms between an island arc and a continental margin.

BANDED-IRON FORMATION A thinly bedded sedimentary rock formed of thin alternating layers of iron oxide and red chert.

BIOTITE A black or brown mica with formula $K(Mg, Fe)_3(AlSi_3O10)(OH)_2$.

BIOTURBATION Mobilization of deposited sediment by the activities of animals or other organisms. May lead to disruption of stratification.

BOLIDE A bright, fiery extraterrestrial body that either collides with Earth or explodes just before contact, creating an impact crater.

BRACHIOPOD Any member of a phylum of solitary marine animals with a bivalve shell and a pair of tentacle-bearing arms for capturing food.

BRECCIA Any coarse-grained rock consisting of angular rock clasts held together by lithified fine-grained matrix or mineral cement.

C

CALICHE A hardened crust of calcium carbonate that often forms over soil in dry regions.

CATASTROPHISM The concept that Earth history is punctuated by occasional sudden and devastating catastrophes.

CELESTITE A pale-blue sulfate mineral with the formula $SrSO_4$.

CHANNELED SCABLANDS Areas with large-scale sedimentary structures (such as giant dunes) indicating massive sediment transport and landscape modification during floods associated with breaching of ice dams in large glacial lakes.

CHERT A chemical sedimentary rock made of quartz composed of such fine crystals that it appears glassy.

CLASTIC SEDIMENTS Sediments derived from mechanical breakdown of preexisting rocks.

CLAY 1. Any particle of sediment less than four microns in diameter. 2. Any member of a class of hydrous aluminum-bearing phyllosilicate minerals.

COLUMNAR JOINTING A joint system in basalt formed by contraction of cooling lava. The joints tend to delineate vertical, hexagonal prisms of rock.

CRATER A funnel-shaped depression formed by bolide impact or by collapse or explosion of a volcanic vent.

CROSSBEDS Inclined, striplike beds of sand, silt, or gravel formed by the erosion and redeposition of sediment that forms ripples or dunes.

CROSS-SECTION A diagram of the features revealed in a geological structure by a vertical plane passing through the structure.

CYCADS A group of seed-forming plants with thick, stubby trunks and fernlike compound leaves.

CYCLOTHEM A rhythmic repetition of strata, made up of several different kinds of distinctive sedimentary rock units that is repeated upward through the stratigraphic succession.

D

DIASTEM A minor break or gap in sedimentation in a succession of sedimentary strata.

DIP Refers to the angle below the horizontal of a tilted stratum or feature.

DISTRIBUTARY A branch of a river flowing away from the main channel and not rejoining it.

DOLOMITE 1. A carbonate mineral with formula $CaMg(CO3)_2$. 2. A carbonate sedimentary rock made largely of the mineral dolomite.

DRUMLIN An elongated hill or ridge formed by the deposition of till.

E

EPICENTER That point on the Earth's surface that is directly above the focus of an earthquake.

EUKARYOTE An organism in which genetic material is located in the nuclei of a complex cell that is surrounded by a membrane.

EVAPORITE A rock layer consisting of highly soluble minerals deposited as a chemical sediment as a result of desiccation of a body of salt water.

F

FAULT A planar rock fracture along which motion has occurred, displacing the sides of the fracture.

FAUNAL SUCCESSION The replacement, in a given habitat or geologic time period, of animals lost to extinction by newly evolved forms.

FELDSPAR An abundant class of tectosilicate minerals bearing potassium, aluminum, sodium, and/or calcium in addition to silica tetrahedra.

FJORD A narrow, steep-sided marine inlet excavated by the action of glaciers.

FORAMINIFERA A type of unicellular protist. Many forms are marine and secrete skeletons (tests) of calcium carbonate. The test has one or more chambers and is typically composed of calcite.

G

GABBRO An ultramafic igneous rock with coarse texture consisting of amphiboles (and other iron and magnesium-rich minerals such as orthopyroxenes) and calcic plagioclase.

GAIA THEORY A theory proposed by James Lovelock, which states that all life on Earth functions as part of a self-regulating system.

GARNET Any member of a group of nesosilicate minerals common in metamorphic rocks such as mica schists and distinguished by compact, isometric crystals.

GAS HYDRATE A type of solid in which gas molecules are trapped within a lattice of hydrogen-bonded water molecules.

GEOPHYSIOLOGY The study of the interactions between the physical and biological parts of a planet, with special focus on feedback mechanisms and regulation of climate.

GLACIAL ERRATIC A rock block, often of exotic lithology in comparison to the underlying terrane, transported and deposited by the motion of glacial ice.

GONDWANA An ancient supercontinent that is believed to have contained most of the landmasses that are now located in the Southern Hemisphere.

GRABEN A valley formed by parallel normal faults dipping toward each other.

GRADED BED A sedimentary layer in which the coarsest grains are at the base of the bed and the finer grains above.

GRADUALISM The concept that major change occurs as the incremental summation of numerous small changes over time.

GRAYWACKE A dark colored sandstone composed of angular/subangular quartz grains, feldspar grains, and lithic fragments embedded in a fine-grained matrix.

GREAT RIFT VALLEY A geological fault system of southwest Asia and east Africa; it extends about 3,000 miles (4,830 km) from northern Syria to central Mozambique.

GRENVILLE OROGENY An approximately billion-year-old mountain-building event that resulted in the formation of Rodinia.

GUYOT A flat-topped seamount.

H

HANGING VALLEY A tributary valley with a discordant junction with the trunk valley. The discordance is usually due to rapid downcutting of the main valley due to fluvial erosion or glacial action.

HEMATITE An iron oxide mineral with formula Fe_2O_3. Its color varies from black to red-brown, and it makes a characteristic red color when powdered.

HORST A raised portion of the Earth's crust that is adjacently surrounded by depressed faults.

HYPERSEA Eukaryotic life on land, plus any symbionts or parasites living within the tissues of eukaryotic land organisms, be they other eukaryotes, bacteria, viruses, or any viruslike entities capable of living in and/or reproducing within the tissues of a host.

I

IGNEOUS ROCK Rock formed by cooling of magma.

IGNIMBRITE A volcanic deposit formed by welded volcanic ash.

ISOSTASY As applied in geology, the principle that more-massive slabs of crustal rock will settle deeper into the mantle.

ISOTOPES Atoms of the same element with differing numbers of neutrons in the atomic nucleus.

K

KARST Cavities, caves, and sinkholes on a predominantly limestone terrane formed by the decalcification and leaching of soluble rocks.

L

LAHAR An energetic mudflow of water-saturated volcanic mud.

LAMINA A layer of sediment or bed of sedimentary rock less than one centimeter thick.

LAURASIA An ancient supercontinent that contained most of the landmasses that now make up the Northern Hemisphere.

LEFT-LATERAL FAULT A strike-slip fault in which the side of the fault opposite from the side on which one is standing has been displaced to the left.

LIMESTONE A chemical or bioclastic sedimentary rock composed primarily of calcium carbonate.

LOBOPOD An arthropod-like animal with three or more pairs of unjointed legs on a cylindrical body.

LOESS A deposit of wind-blown glacial silt.

M

MAFIC A type of magma, silicate mineral, or igneous rock that has a high concentration of "heavy" elements such as magnesium and iron.

MAGNETITE A strongly magnetic, black iron oxide mineral with formula Fe_3O_4.

MANTLE The part of the Earth's interior between the core and the crust.

MASS EXTINCTION Extinction of many different types of unrelated organisms during a geologically brief interval of time.

MASS WASTING Occurs when soil, regolith, or rock travel downslope once the gravitational forces working upon a slope become greater than the forces maintaining the slope's stability.

MESONYCHID A group of early Cenozoic predatory animals thought to have been ancestral to cetaceans (whales).

MICRON One one-thousandth of a millimeter.

MID-OCEAN RIDGE An elevated part of the seafloor where new ocean crust is created by upward flow of magma.

MOHOROVICIC DISCONTINUITY The boundary between Earth's crust and mantle. Named for Croatian seismologist Andrija Mohorovicic, it represents the level at which P-wave velocities increase from approximately 7 km/sec to 8 km/sec.

MORAINE rock and soil debris that is stripped from a mountain, transported, and eventually deposited by glacial movement.

N

NODULE An irregular mass of rock, with a composition differing from the surrounding rock, that forms by precipitation within the host rock.

NUNATAK An exposed mountain or mountainous region projecting above a glacial ice sheet.

O

OBDUCTION The process by which large-scale thrust faults superimpose dense seafloor crust over buoyant continental crust.

OBSIDIAN Volcanic glass, typically dark in color, with a characteristic conchoidal fracture.

OCEANIC CRUST The mafic crust of the oceans, with a thickness averaging about three miles (5 km).

OOID A spherical grain of sand composed of concentrically precipitated layers of calcium carbonate or other mineral.

OOLITE A carbonate rock composed of ooids.

OORT CLOUD A diffuse swarm of comets and other bolides at the outer margins of the solar system.

P

PALEOSOL Ancient soils buried underneath newer sedimentary or volcanic deposits.

PALYNOLOGY The study of microfossils (pollen, spores, protist cysts) consisting of organic compounds such as sporopollenin.

PANGEA A Late Paleozoic and Early Mesozoic supercontinent consisting of virtually all the continents.

PARAGENESIS The sequence of deposition or formation of different types of minerals in ores and metamorphic rocks.

PERMINERALIZATION Addition, during fossilization, of mineral deposits (from ground or sediment pore water) to organic remains being fossilized.

PHYLLOSILICATES Silicate minerals consisting of sheets of silica tetraheda. There is perfect cleavage in phyllosilicate minerals parallel to the orientation of these sheets.

PILLOW BASALT A volcanic rock deposit consisting of stacked, saclike structures with chilled (glassy) rims formed by the eruption of lava under water.

PLACER A stream or river deposit of concentrated heavy minerals such as gold.

PLAYA An intermittent desert lake with no outlet. Commonly associated with the deposition of evaporite minerals.

PROTOLITH A rock, such as limestone, that has not undergone metamorphism.

PLUTONIC ROCK Rock formed by cooling of magma below Earth's surface.

POINT BAR The curved shoreline on the inner edge of a meander loop, and parallel ridgelike deposits inland from this shore. Formed by successive deposits of riverine sediment.

POLYNYA A large area of water surrounded by sea ice.

PROVENANCE The region or source rock from which a particular sediment or sedimentary clast is derived.

PUMICE A silica-rich glassy volcanic rock so filled with gas bubbles that it floats on water.

P-WAVE The primary or pressure wave of an earthquake. It is the first wave to arrive in an episode of earthquake motion. Unlike S-waves, P-waves will travel through liquids.

PYRITE The most common sulfide mineral, with formula FeS_2. It has a distinctive brassy yellow color.

PYROXENE Any member of a group of inosilicate minerals with simple chains of silica tetrahedra in the crystal structure. The ratio of silicon to oxygen in pyroxenes is one to three.

R

REGOLITH A layer of unconsolidated, heterogeneous material, or "blanket rock," that covers solid rock.

REGRESSION Withdrawal of seawater from land surface; lowering of sea level.

RHYOLITE A fine-grained igneous rock with a mineralogy similar to that of granite.

RIFTING The tectonic sundering of a continent or supercontinent into one or more smaller continental fragments.

RODINIA A supercontinent consisting of all or nearly all of the continents. Consolidated one billion years ago, this supercontinent broke into smaller continents well before the Cambrian.

S

SCIORRUCK A submarine landslide that triggers a tsunami. Such events are capable of battering coastal areas with little warning.

SHIELD VOLCANO A volcano that is characterized by sides with shallow, gradual slopes, and eruptions of low-viscosity lava.

SILTSTONE A clastic sedimentary rock composed of sediment grains ranging in diameter between 63 and 64 microns (also called silt).

SKARN A metamorphic deposit in which a limestone or other carbonate rock has been intruded by a silicate magma or silicate-rich hydrothermal fluid.

SMITHSONITE A carbonate mineral with formula $ZnCO_3$. An ore of zinc usually encountered in limestones.

SNOWBALL EARTH An extreme phase of glaciation in which glaciers reach tropical latitudes.

SOLE The undersurface of a bed.

STRATIGRAPHY The branch of geology dealing with the chronology and succession of sedimentary or other types of stratified rocks. Also, the study of the age, accumulation, and depositional environments of ancient strata.

STRIKE The direction of the line formed by the intersection of a bedding plane with a horizontal plane tangent to the Earth's surface.

STRIKE-SLIP FAULT A type of fault that is characterized by a near-vertical surface, lateral movement of the footwall, and very little vertical motion.

STROMATOLITE A finely laminated organo-sedimentary structure thought to have been formed by the sediment-trapping, -binding, and -secreting activities of bacteria forming a seafloor bacterial mat.

SUBDUCTION ZONE A zone where oceanic crust flexes downward and dives into the mantle, underneath either continental crust or other oceanic crust.

SYMBIOGENESIS The evolutionary creation of a new life-form by the combination in symbiosis of two or more, once free-living, unrelated species of organisms.

T

TECTONICS The study of a planet's (or moon's) crust and the forces that operate upon it.

TEPHRA Any airborne material, including ash, that is produced as a result of a volcanic eruption.

TERRANE An area or surface that is defined by a prevalence of a particular type of rock or rock group.

TILL Unstratified glacial debris that consists largely of clay, sand, and gravel.

TRACE FOSSILS Tracks, trails, burrows, fecal pellets, or other sediment disturbances caused by the activities of animals or other types of organisms.

TRANSFORM FAULT A vertical fault formed at, and perpendicular to, the margin of tectonic plates that are sliding past one another.

TRANSGRESSION Landward motion of the shoreline.

TRAVERTINE A calcitic or aragonitic, porous cave deposit present as a heavy and hard variety calcareous tufa.

TUFA Carbonate or siliceous deposits formed in the vicinity of springs, hot springs, or lakeshores. Pronounced TOO-fah.

TUFF Material erupted from a volcano or volcanic vent that solidifies in the air before being deposited. It thus includes volcanic particles ranging is size from volcanic ash to volcanic bombs.

TURBIDITE Thinly bedded deposits, often with an organized sequence of stratification (Bouma sequence), formed by the settling and deposition of turbidity currents in deep water areas.

U

UNIFORMITARIANISM A scientific philosophy, subscribed to by Charles Lyell, that proposes that the forces and laws that govern the universe in the present have always operated in the same manner and will continue to do so forever.

UREY REACTION The chemical reaction $CO_2 + CaSiO_3 => CaCO_3 + SiO_2$. It is important in climate studies as it specifies the absorption of the greenhouse gas carbon dioxide by the chemical weathering of a silicate mineral such as wollastonite.

U-SHAPED VALLEY A valley, originally V-shaped, that has been broadened by glacial erosion.

V

VARVE A set of paired layers in a clay deposited in a glacial lake. The couplet consist of a summer layer (coarser grained, including sand and silt mobilized during spring melt) and a winter layer (finer grained deposited in quiet water conditions under the ice).

WOLLASTONITE A mineral with formula $CaSiO_3$. Common in metamorphosed limestones.

X

XENOLITH An isolated foreign inclusion in a plutonic igneous rock that has a composition and color different from that of the surrounding rock.

Y

YOUNGER DRYAS An episode during the end (12,000 years ago) of the most recent glaciation characterized by a brief return to cold conditions.

Z

ZIRCON A mineral with formula $ZrSiO_4$. Found in all types of igneous rocks and in many sandstones, zircon is important for radiometric dating because of the presence in this crystal of small amounts of uranium and thorium.

ZONING Concentric growth layers of a crystal that record changes in the mineral composition as one goes from the center of the crystal to its outer, and most recently formed, layer.

FURTHER READING

BOOKS

Allen, K. C., and D. E. G. Briggs. *Evolution and the Fossil Record.* London: Belhaven Press, 1989.

Alvarez, W. *T. rex and the Crater of Doom.* New York: Vintage Press, 1998.

Benton, M. J. *When Life Nearly Died: The Greatest Mass Extinction of All Time.* London: Thames & Hudson, 2003.

Beus, S. S., and M. Morales. *Grand Canyon Geology.* Oxford: Oxford University Press, 2002.

Bjornerud, M. *Reading the Rocks: The Autobiography of the Earth.* Cambridge, Massachusetts: Westview Press, 2005.

Boggs, S. *Principles of Sedimentology and Stratigraphy*, 4th ed. Englewood, New Jersey: Prentice Hall, 2005.

Bucher, K., and M. Frey. *Petrogenesis of Metamorphic Rocks.* New York: Springer, 2002.

Catuneau, O. *Principles of Sequence Stratigraphy.* Amsterdam: Elsevier Academic Press, 2006.

Clarkson, E. N. K. *Invertebrate Palaeontology and Evolution*, 4th ed. Oxford: Blackwell Science, 1998.

Cloud, P. *Cosmos, Earth and Man: A Short History of the Universe.* New Haven, Connecticut: Yale University Press, 1978.

———. *Oasis in Space: Earth History form the Beginning.* New York: W. W. Norton, 1988.

Coenraads, R. R. *Rocks and Fossils: A Visual Guide.* Richmond Hill, Ontario: Firefly Books, 2005.

Condie, K. C. *Earth as an Evolving Planetary System.* Amsterdam: Elsevier Academic Press, 2004.

Conway Morris, S. *The Crucible of Creation: The Burgess Shale and the Rise of Animals.* Oxford: Oxford University Press, 1998.

———. *Life's Solution: Inevitable Humans in a Lonely Universe.* Cambridge: Cambridge University Press, 2004.

Cook, T., and L. Abbott. *Hiking the Grand Canyon's Geology.* Seattle, Washington: Mountaineers Books, 2004.

Cvancara, A. M. *Sleuthing Fossils: The Art of Investigating Past Life.* New York: Wiley, 1990.

Ellis, R. *Aquagenesis.* New York: Penguin, 2003.

Erwin, D. H. *Extinction: How Life on Earth Nearly Ended 250 Million Years Ago.* Princeton, New Jersey: Princeton University Press, 2006.

Feldman, J. *When the Mississippi Ran Backwards: Empire, Intrigue, Murder, and the New Madrid Earthquakes.* New York: Free Press, 2005.

Fortey, R. *Earth: An Intimate History.* New York: Knopf, 2004.

Fowler, C. M. R. *The Solid Earth: An Introduction to Global Geophysics*, 2nd ed. Cambridge: Cambridge University Press, 2004.

Fritz, W. J., and J. N. Moore. *Basics of Physical Stratigraphy and Sedimentology.* New York: Wiley, 1988.

Fry, I. *The Emergence of Life on Earth: A Historical and Scientific Overview.* New Brunswick, New Jersey: Rutgers University Press, 2000.

Gains, R. V., H. C. W. Skinner, E. E. Foord, B. Mason, and A. Rosenzweig. *Dana's New Mineralogy*, 8th ed. New York: Wiley, 1997.

Glaessner, M. F. *The Dawn of Animal Life: A Biohistorical Study*. Cambridge: Cambridge University Press, 1985.

Gould, S. J. *The Structure of Evolutionary Theory*. New York: Belknap Press, 2002.

Hallam, A. *Great Geological Controversies*, 2nd ed. Oxford: Oxford University Press, 1992.

Hallam, A., ed. *Atlas of Palaeobiogeography*. Amsterdam: Elsevier Scientific Publishing, 1973.

Hallam, A., and P. Wignall. *Mass Extinctions and their Aftermath*. Oxford: Oxford University Press, 2003.

Halstead, L. B. *The Search for the Past: Fossils, Rocks, Tracks and Trails, the Search for the Origin of Life*. Garden City, New York: Doubleday, 1982.

Harland, D. M. *Water and the Search for Life on Mars*. New York: Springer Praxis Books, 2005.

Hartmann, W. K. *A Traveler's Guide to Mars: The Mysterious Landscapes of the Red Planet*. New York: Workman Press, 2003.

Hoyt, W. G. *Coon Mountain Controversies: Meteor Crater and the Development of Impact Theory*. Tucson, Arizona: University of Arizona Press, 1987.

Jenkins, G. S., M. A. S. McMenamin, C. P. McKay, and L. Sohl, eds. *The Extreme Proterozoic: Geology, Geochemistry, and Climate*. Geophysical Monograph 146. Washington, D.C: American Geophysical Union.

Jenny, H. *Factors of Soil Formation: A System of Quantitative Pedology*. Mineola, New York: Dover Publications, 1994.

Knoll, A. H. *Life on a Young Planet: The First Three Billion Years of Evolution on Earth*. Princeton, New Jersey: Princeton University Press, 2004.

Koene, C. J. *"The Chemical Constitution of the Atmosphere from Earth's Origin to the Present, and its Implications for Protection of Industry and Ensuring Environmental Quality"* (18560. Translated and edited by Mark A. S. McMenamin. New York: Mellen Press, 2004.

Kunzig, R. *Mapping the Earth: The Extraordinary Story of Ocean Science*. New York: Norton, 2000.

Lahav, N. *Biogenesis: Theories of Life's Origin*. New York: Oxford University Press, 1999.

Lane, N. *Oxygen: The Molecule That Made the World*. Oxford: Oxford University Press, 2002.

Lockley, M. *The Eternal Trail: A Tracker Looks at Evolution*. Reading, Massachusetts: Perseus Books, 1999.

Lopes, R. M. C., and T. K. P. Gregg. *Volcanic Worlds: Exploring the Solar System's Volcanoes*. Berlin: Springer, 2004.

Love, J. C. "In Memory of Christina Lochman-Balk, 1907–2007." *New Mexico Geology*, vol. 28, n. 3, pp. 88–90. 2006.

Lowenstam, H. A., and S. Weiner. *On Biomineralization*. Oxford: Oxford University Press, 1989.

Lunine, J. I., and C. J. Lunine. *Earth: Evolution of a Habitable World*. Cambridge: Cambridge University Press, 1998.

Margulis, L. *Symbiosis in Cell Evolution*. New York: Freeman, 1993.

———. *Symbiotic Planet: A New Look at Evolution*. New York: Basic Books, 2000.

Margulis, L., C. Matthews, and A. Haselton, eds. *Environmental Evolution: Effects of the Origin and Evolution of Life on Planet Earth*. Cambridge, Massachusetts: The MIT Press, 2000.

Martin, Anthony J. *Introduction to the Study of Dinosaurs,* 2nd ed. Malden, Massachusetts: Blackwell Publishing, 2006.

Marvin, N., and J. James. *Chased by Sea Monsters: Prehistoric Predators of the Deep.* New York: DK Publishing, 2004.

Mathews, D. *Rocky Mountain Natural History: Grand Teton to Jasper.* Raven Editions, 2003.

McCarthy, T., and B. Rubidge. *The Story of Earth and Life: A Southern African Perspective on a 4.6-Billion-Year Journey.* Cape Town: Struik Publishers, 2006.

McDonald, N. G. *The Connecticut Valley in the Age of Dinosaurs: A Guide to the Geologic Literature, 1681–1995.* Bulletin 116. Hartford: State Geological and Natural History Survey of Connecticut, 1996.

McMenamin, M. A. S. *The Garden of Ediacara: Discovering the First Complex Life.* New York: Columbia University Press, 1998.

———. *Dictionary of Earth and Environment.* South Hadley, Massachusetts: Meanma Press, 2001.

McMenamin, M. A. S., and D. L. S. McMenamin. *The Emergence of Animals: The Cambrian Breakthrough.* New York: Columbia University Press, 1990.

———. *Hypersea: Life on Land.* New York: Columbia University Press, 1994.

McPhee, J. *Annals of the Former World.* New York: Farrar, Straus and Giroux, 2000.

———. *Basin and Range.* New York: Farrar, Straus and Giroux, 1982.

Miall, A. D. *The Geology of Fluvial Deposits: Sedimentary Facies, Basin Analysis, and Petroleum Geology.* New York: Springer, 2006.

Neaverson, E. *Stratigraphical Palaeontology: A Manual for Students and Field Geologists.* London: Macmillan, 1928.

Parker, A. *In the Blink of an Eye: How Vision Sparked the Big Bang of Evolution.* New York: Basic Books, 2004.

Prothero, D. R., and R. H. Dott, Jr. *Evolution of the Earth,* 7th ed. Boston: McGraw-Hill, 2004.

Redfern, R. *Origins: The Evolution of Continents, Oceans and Life.* Norman: University of Oklahoma Press, 2001.

Retallack, G. J. *Soils of the Past.* Blackwell Publishing, 2001.

Rudwick, M. J. S. *The Great Devonian Controversy.* Chicago: University of Chicago Press, 1985.

Savoy, L. E., E. M. Moores, and J. E. Moores. *Bedrock: Writers on the Wonders of Geology.* San Antonio, Texas: Trinity University Press, 2006.

Schneider, S. H., J. R. Miller, E. Crist, and P. J. Boston, eds. *Scientists Debate Gaia: The Next Century.* Cambridge, Massachusetts: The MIT Press, 2004.

Schopf, J. W. *Cradle of Life: The Discovery of Earth's Earliest Fossils.* Princeton, New Jersey: Princeton University Press, 2001.

Schwartzman, D. *Life, Temperature, and the Earth: The Self-Organized Biosphere.* New York: Columbia University Press, 1999.

Seilacher, A. *Fossil Art.* Royal Tyrrell Museum of Palaeontology; Alberta, Canada: Drumheller, 1997.

———. *Trace Fossil Analysis.* New York: Springer, 2007.

Selley, R. C., R. M. Cocks, and I. R. Plimer, eds. *Encyclopedia of Geology.* Oxford: Elsevier, 2004.

Shubnikov, A. V., and N. N. Sheftal. *Growth of Crystals,* Vol. 3. New York: Consultants Bureau, 1962.

Skehan, J. W. *Roadside Geology of Massachusetts.* Missoula, Montana: Mountain Press Publishing, 2001.

———. *Geology and Grace: Teilhard's Life and Achievements.* Teilhard Studies Number 53. New York: American Teilhard Association, 2006.

Stanley, S. M. *Children of the Ice Age: How a Global Catastrophe Allowed Humans to Evolve.* New York: Freeman, 1998.

Valentine, J. W. *On the Origin of Phyla.* Chicago: University of Chicago Press, 2004.

Vernadsky, V. I. *The Biosphere: Complete Annotated Edition.* New York: Copernicus, 1998.

Volk, T. *Gaia's Body: Toward a Physiology of Earth.* New York: Springer-Verlag, 1998.

Walker, G. *Snowball Earth: The Story of a Maverick Scientist and His Theory of Global Catastrophe That Spawned Life as We Know It.* New York: Three Rivers Press, 2004.

Ward, P., and D. Brownlee. *Rare Earth: Why Complex Life Is Uncommon in the Universe.* New York: Springer, 2003.

Whiteley, T. E., G. J. Kloc, and C. E. Brett. *Trilobites of New York: An Illustrated Guide.* Ithaca, New York: Cornell University Press, 2002.

Whybrow, P. J., and A. Hill. *Fossil Vertebrates of Arabia.* New Haven, Yale University Press, 1999.

Willis, K. J., and J. C. McElwain. *The Evolution of Plants.* Oxford: Oxford University Press, 2002.

Wilson, E. O. *The Diversity of Life.* New York: Norton, 1999.

Winchester, S. *The Map That Changed the World: William Smith and the Birth of Modern Geology.* New York: HarperCollins, 2001.

———. *A Crack in the Edge of the World: America and the Great California Earthquake of 1906.* New York: HarperCollins, 2005.

———. *Krakatoa: The Day the World Exploded: August 27, 1883.* New York: HarperCollins, 2005.

Wood, D. *Five Billion Years of Global Change: A History of the Land.* New York: Guilford Press, 2003.

Yochelson, E. L. *Smithsonian Institution Secretary, Charles Doolittle Walcott.* Kent, Ohio: Kent State University Press, 2001.

WEB SITES

Animations of Plate Tectonics
csep10.phys.utk.edu/astr161/lect/earth/tectonics.html

Arches National Park
www.desertusa.com/arches/index.html

Author Profile
www.mtholyoke.edu/acad/earth/profiles/mcmenamin.shtml

Big Bend, Texas
geowww.geo.tcu.edu/bigbend/p16.html

Columbia Earthscape, an Online Resource on the Global Environment
www.earthscape.org/

Comet sample return mission
civspace.jhuapl.edu/

Cuesta definition
en.wikipedia.org/wiki/Cuesta

Dinosaur paleontology and expeditions
www.dinoruss.com/

Earth Education online
earthednet.org/

Ediacaran fossils
geol.queensu.ca/people/narbonne/recent_pubs1.html

Elastic rebound animation
projects.crustal.ucsb.edu/understanding/elastic/intro-rebound.html

Exploration of Mars
mars.jpl.nasa.gov/missions/

Fault motion animations
www.iris.edu/gifs/animations/faults.htm

Galileo Jupiter project
www2.jpl.nasa.gov/galileo/

Genesis Discovery 5 mission
www.gps.caltech.edu/genesis/

Geological Maps and More
oddens.geog.uu.nl/index.php

Geology software
www.geologynet.com/

Geophysical equipment source
www.giscogeo.com/

Global Ocean Data Analysis Project
cdiac.ornl.gov/oceans/GLODAP/GlopOV.htm

Gondwana breakup
www.scotese.com/satlanim.htm

Google Earth
earth.google.com/

La Brea Tar Pits
www.tarpits.org/

Lunar and Planetary Institute
www.lpi.usra.edu/

Lunar prospecting
www.nasa.gov/audience/foreducators/Redirect_
Spacelink.html

Martian minerals
www.mtholyoke.edu/courses/mdyar/marsmins/

Mineral and Rock collecting
www.rockhounds.com/

Mineral exploration
www.amebc.ca/

Minerals in Thin Section
www.gly.bris.ac.uk/www/teach/opmin/mins.html

Mineralogy Database (rotating graphics!)
webmineral.com/

Mining technology
www.infomine.com/

Museum of the Earth
www.museumoftheearth.org/

NOAA Center for Tsunami Research
nctr.pmel.noaa.gov/

North Carolina Museum of Natural History
www.dinoheart.org/

Palaeos: The Trace of Life on Earth
www.palaeos.com/

Clockwise from top: Mica, galena, galcite, and ghalcopyrite.

Paleobiology database
paleodb.org/cgi-bin/bridge.pl

Paleomap project by Christopher R. Scotese
www.scotese.com/info.htm; http://www.scotese.com/
newpage13.htm

Petrology course
www.eos.ubc.ca/courses/eosc221/index.html

Reelfoot Rift and its earthquakes
quake.ualr.edu/public/reelfoot.htm

Rockware Software source
www.rockware.com/

Search professional papers
scholar.google.com/

Seismic waves/eruption software
www.geol.binghamton.edu/faculty/jones/

Smithsonian research
www.smithsonian.org/research/

Snowball Earth Web site
www.snowballearth.org/

Spatial Data Processing
www.caris.com/

Stone Forest Karst Field Trip, Kunming, China
www.uh.edu/~jbutler/kunming/stoneforestkunming.
html

Sue the female Tyrannosaurus rex at the Field
Museum, Chicago
www.fieldmuseum.org/sue/index.html

Surviving Snowball Earth
www.livescience.com/animalworld/060607_snowball_
earth.html

Tectonic maps of Rodinia
www.tsrc.uwa.edu.au/440project/rodiniamaps

The Disaster Center
www.disastercenter.com/

Tutorial on Remote Sensing
rst.gsfc.nasa.gov/

Understanding Plate Motion
pubs.usgs.gov/gip/dynamic/understanding.html

United States Geological Survey
www.usgs.gov/

Uranium mining
www.rodiniaminerals.com/

US Geological Survey Earthquake Hazards Program
earthquake.usgs.gov/

Virtual Fossils
www.nhm.ac.uk/nature-online/virtual-wonders/

Virtual Geology Field Trips
www.uh.edu/~jbutler/anon/quick.html

Volcano Information Center at the University of
California at Santa Barbara
volcanology.geol.ucsb.edu/

Wellsite Geological Supplies
www.usgeosupply.com/

Wind River virtual field trip
www.wind-river.com/Multimedia/QTVR/

World-Wide Earthquake Locater, Edinburgh Earth
Observatory
tsunami.geo.ed.ac.uk/local-bin/quakes/mapscript/
home.pl

AT THE SMITHSONIAN

If you wish to learn more about the scientific processes that helped to shape our world, there is perhaps no better place to visit than the National Museum of Natural History in Washington, D.C. Since it opened in 1910, the museum has become a true warehouse of scientific knowledge as well as one of the preeminent centers for geological study.

Much of the Smithsonian Institution's geological material can be found in the Janet Annenberg Hooker Hall of Geology, Gems and Minerals. The hall boasts one of the most impressive collections of minerals and gems in the world and has more than 2,500 objects on display. It is also the permanent home of one of the most sought-after gems in the world, the legendary Hope Diamond. In addition to the mineral and gem displays, the hall also contains re-creations of mines, as well as galleries dedicated to the study of plate tectonics and volcanic activity.

The Smithsonian Institution has also created an on-line presentation dedicated to the study of geology. The Dynamic Earth is a Web-based multimedia experience that covers such topics as rock formation, tectonic theory, and the geology of our solar system. The site can be accessed at www.mnh.si.edu/earth/main_frames.html. The Smithsonian

The golden-domed National Museum of Natural History is located on the north side of the National Mall in Washington, D.C. Scholars from around the world regularly visit the valuable collections and resources relating to natural history.

The mineral smithsonite is named for James Smithson, founder of the Smithsonian Institution.

is also in the process of creating a virtual tour of the Janet Annenberg Hooker Hall, which should be finished by early 2007.

Smithsonian founder James Smithson was a trained mineralogist with an extensive collection of meteorites, and his studies have had a lasting impact on the museum's collections. Although Smithson's early meteorite samples were lost in a fire, the Smithsonian Institution is now in possession of one of the most complete collections of meteorites in the world. The U.S. National Meteorite Collection, which was founded in 1870, is also located in the National Museum of Natural History. The collection includes pieces of every known type of meteorite and boasts a total of more than 9,250 distinct meteorite specimens. The collection also includes just under 7,000 thinly sliced meteorite sections. Scientists from all over the world come to study these samples in the hopes that geological assessment of the rock composition will allow some insight into the processes that formed not only this planet, but the entire solar system.

The Smithsonian Institution is dedicated to increasing knowledge of nature. Scientists at the Department of Mineral Sciences at the National Museum of Natural History study the origin and evolution of the Earth and solar system, geological processes, and the effects of geologic and meteoritic phenomena on Earth's atmosphere and biosphere. At the Center for Earth and Planetary Studies, a research unit within the National Air and Space Museum, scientists perform research on planetary science, terrestrial geophysics, and the remote sensing of environmental change.

Thanks to the Smithsonian Traveling Exhibition Services (SITES), you do not have to go to Washington to experience the thrill of the Smithsonian Institute. For over 50 years, SITES has been bringing materials of the Smithsonian to museums and schools all over the country. Currently, SITES is promoting Caves: A Fragile Wilderness, an exhibition that explores the delicate ecosystems of our planet's cave systems. Consisting of 39 color photographs and accom-

panying text, A Fragile Wilderness provides us with an up-close look at various cave systems from Alaska to Malaysia, and reminds us of the myriad threats to these geological marvels. This exhibition is slated to run through April of 2008. For more information about the SITES program, visit: www.sites.si.edu.

Almost all of the specimens that are on display in these exhibits are part of the Smithsonian Rock and Ore collections. The collections are divided into subdivisions, such as the Sea Floor Rock Collection, the Ultramafic Xenolith Collection, and the Island Rocks Collection. A large part of the collection is made up of volcanic rocks. The Smithsonian has organized data and photos from many active and dormant volcanoes as part of its Global Volcanism Project. Visitors to the GVP Web site are able to search for volcanoes by region, type, and other subfeatures. Currently, there are more than 1,546 volcanoes included in the database, making this one of the most inclusive frameworks of volcanic activity that is currently accessible to the public.

The Smithsonian's Rock and Ore collections also serve as an invaluable resource to geologists who can search for specimens by lithology, locality, and chemical analysis among other criteria. These collections contain about 265,000 catalogued samples and another 50,000 specimens that are awaiting accession, including the Shoemaker impactites, Boyd and Wilshire xenoliths, and the Bateman granites.

The National Museum of Natural History is located on Tenth Street and Constitution Avenue NW in Washington, D.C. The museum is open seven days a week from 10 AM to 5:30 PM, and admission is free. For more information about the National Museum of Natural History or the other Smithsonian museums, please send an e-mail to info@si.edu or call (202) 633-1000.

Top: "Persian Gulf" by Peter Jones is one of the 39 photographs that make up Caves: A Fragile Wilderness, one of the current SITES exhibitions. Bottom: The 45.52-carat Hope Diamond, the world's largest deep-blue diamond, is on display in the Jane Annenberg Hooker Hall of Geology, Gems and Minerals. Opposite page: The Janet Annenberg Hooker Hall opened in 1997 and now houses one of the largest collections of gems and minerals in the world.

INDEX

ACKNOWLEDGMENTS & PICTURE CREDITS

First, I would like thank my editor, Lisa Purcell, at Hylas Publishing, who guided this project with considerable insight and energy. Second, I would like to thank Dianna L. Schulte McMenamin, who I can always count on for a key insight, a critical observation, and to keep me in line if I get overbearing during a Monopoly game. Finally, Flight like to thank all of my geology colleagues at Mount Holyoke College and beyond, plus my former professors in geology, particularly Stanley M. Awramik, Jack D. Beuthin, James R. Boles, John C. Crowell, William R. Dickinson, Donald W. Hyndman, James C. Ingle, Keith A. Kvenvolden, Joseph L. Kirchvink, Lynn Margulis, Donald J. Marshall, Dolf Seilacher, James W. Skehan SJ, and James W. Valentine.

The author and publisher also offer thanks to those closely involved in the creation of this volume: Andrew Johnston, Center for Earth and Planetary Studies, National Air and Space Museum; consultant Ross Secord, University of Michigan; Ellen Nanney, Senior Brand Manager, Katie Mann, and Carolyn Gleason with Smithsonian Business Ventures; Collins Reference executive editor Donna Sanzone, editor Lisa Hacken, and editorial assistant Stephanie Meyers; Hydra Publishing president Sean Moore, publishing director Karen Prince, senior editor/designer Lisa Purcell, art director Brian MacMullen, designers Erika Lubowicki and Ken Crossland, production editors Eunho Lee and Anthony Galante, editorial director Aaron Murray, editors Michael Smith, Suzanne Lander, and Rachael Lanicci, picture researcher Ben DeWalt, proofreader Glenn Novak, and indexer Jessie Shiers.

PICTURE CREDITS

The following abbreviations are used:
ALWI–Alfred Lothar Wegener Institute; AP–Associated Press; ARS–Agricultural Research Service; BS–Big Stock Photos; CMGW–Commission for the Geological Map of the World; DLR–German Aerospace Center; EPA–Environmental Protection Agency; ESA–European Space Agency; ESO–European Southern Observatory; FU–Free University of Berlin; GSFC–Goddard Space Flight Center; GSWA–Geological Survey of Western Australia; HST–Hubble Space Telescope; IO–Index Open; IS–iStockphoto.com; JPL–Jet Propulsion Laboratory; LHL–Linda Hall Library; LoC–Library of Congress; MF–Morguefile.com; NASA–National Aeronautics and Space Administration; NMNH–National Museum of Natural History; NOAA–National Oceanic and Atmospheric Association; NRCS–National Resources Conservation Service; NSF–National Science Foundation; NWS–National Weather Service; NYPL–New York Public Library; PD–Public Domain; PNNL–Pacific Northwest National Laboratory; PR–Photo Researchers, Inc.; SIBL–Science Industry and Business Library; SI–Smithsonian Institute; SPL–Science Photo Library; SS–Shutterstock; USDA–United States Department of Agriculture; USDI–United States Department of the Interior; USGS–United States Geologic Survey; USMC–United States Marine Corps.

(t=top; b=bottom; l=left; r=right; c=center)

Front Matter:
iv clipart.com v clipart.com vi IO/Everett Johnson 1t IO/Photolibrary.com 1b clipart.com 2 IO/Wallace Garrison 3t SS/Svetlana Privezentseva 3b clipart.com

Chapter 1: Earth the Planet
4 SPL/David A. Hardy. 5t SS/Stephen Coburn 5b IS/Amanda Rhode 6tl NASA/Don Davis 6r PD 6bl NASA 7 IS/Manik Rattan 8tl SS/Sebastian Kaulitzki 8l NMNH/ Chip Clark 8r NMNH/Chip Clark 9tr NASA/JPL-Caltech/ 9br IS/Brandon Alms 10tl PD 10bl Lisa Purcell 11 SPL/Eye of Science 12tl USDA/ARS/National Soil/David Laird 12b NOAA 13 GSWA

Chapter 2: Planetary Geology
14 Photos.com 15t NASA 15b NASA/JPL/Space Science Institute 16a NASA 16l NASA/GSFC 16br NASA/JPL 17t NASA/JPL 17r NASA/JPL/ Northwestern University 18tl NASA 18bl NASA/ L. Esposito 18br NASA 19bl NASA/JPL 19r NASA/JPL 20tl NASA/JPL/USGS 20b NASA 21t NASA 21b ESA/DLR/FU/G. Nekum 22tl NASA 22tr NASA/ ESA 23 NASA 24tl NASA 24bl SS/Stephen Girimont 24c NASA/JPL/Space Science Institute 25tl NASA 25br

University of Arizona/NASA 26tl NASA 26br NASA 27tl R. Evans/J. Trauger/ H. Hammel/HST Comet Science Team and NASA 27br NASA/JPL 28l NWS 28r University of Wisconsin/Lawrence Sromovsky 28bl JPL 29t NASA 29br NASA/ESA/ESO Space Telescope European Facility/Dr. R. Albrecht

Chapter 3: Minerals, Rocks, and the Crust
30 NOAA 31t USGS/Cascades Volcano Observatory 31b NMNH 32tl SS/Guojón Eyjólfur Ólafsson 32tc NASA 32bl photos.com 33t MS Book and Mineral Company 33b University of North Carolina-Wilmington 34tl Wikipedia 34r NOAA/University of Washington 34bl Shutterstock/ Bryan Busovicki 35tl USGS/ J.Lowenstern 35bl Drexel University 36tl USGS 36l Chris Bolger 36br Drexel University 37tl Wikipedia 37tr IS/Therese McKeon 38tl Drexel University 38tr Drexel University 38l Drexel University 39 Drexel University 40tl SS/Mark Scott 40l Drexel University 40br Drexel University 41tl Drexel University 41tr Hooper Natural History Museum 42tl Wikipedia/Tom Pfeiffer 42r USGS/Austin Post 43t Photos.com 43b Associated Press 44tl USGS 44c Drexel University 45tl Drexel University 45br Drexel University

Chapter 4: Weathering and Soils
46 Photos.com 47t USDA/Nature Source/PR 47b Wikipedia/Christian Fischer 48tl USGS 48b Wikipedia/Eurico Zimbres 49t Kurt Hollocher/Union College 49b Drexel University 50tl Martin Ruzek/USRA 50b Ron Amundson/University of California 51tr Pleum Chenaphun 51b IS/ Malcolm Romain 52tl Photos.com 52l USGS 52r Stan Celestian/Glendale Community College 53tr Weblogs 53bl SPL/Jerry Mason 54tl Photos.com 54l SS/3poD Animation 54r IS/Mark Rasmussen 55tl SS/Sergey Chushkin 55br USGS

Chapter 5: A Living Planet
56 Photos.com 57t MF/digiology 57b NOAA 58tl Yellowstone National Park Service 58bl Wikimedia 59 SPL/ Dirk Wiersma 60tl NOAA 60b NOAA 61tc USGS/ The Malik Project 61r PNNL 62tl WVU/Acadaweb 62b Pleum Chenaphun 63 PR/Eye of Science 64tl morguefile/d0g3 64r PD 65tr Wikimedia 65br morguefile/Rhaeel 66tl photos.com 66–67 PR/Sheila Terry 68tl NASA 68l USDA/FAS 69 Pleum Chenaphun

Chapter 6: The Fossil Record
70 SS/Alexey Krychokov 71t David C. Ward/ Wikipedia 71b SS/Ismael Montero Verdu 72tl USDA/ De Wood, color by Chris Pooley 72b MF/Clarita 73tl Scripps Institution of Oceanography 73br Jon Zander/ Wikipedia 74tl PD 74b NASA 75tr SS/Ismael Montero Verdu 75bl SS/Michael Ledray 76tl NASA 76r NOAA 76bl PD/ Michael Abbey 77 NOAA 78TL SS/Kim Worrell 78tl IS/Zeliha Gurkan 78br IS/Heather Cash 79 SS/Ismael Montero Verdu 80tl PD 80b PR/M.I. Walker 81tr NOAA/ James McVey 81br Photos.com 82tl PD 82c PR/NLM 83t Pleum Chenaphun 83b BS/Mark Hangrove 84tl SS Christine Nichols 84r PR 84b PR/Chase Studio 85 PR/Peter Scoones 86tl SS/Ismael Montero Verdu 86b IS/Keoni Mahelona 87t Eye of Science/SPL 87b SPL/Martin Land 88tl SS/Nicola Keegan 88r SPL/George Bernard 88b MF/Bob Ainsworth 89r SPL/D. Van Ravensway 89b Pleum Chenaphun 90b photos.com 90l PR/Francois Gohier 90–91 photos.com 92tl IS/Adrian Chesterman 92bl SS/Bob Ainsworth 92c SS/Ismael Montero Verdu 93 SS/John Kirinic

Ready Reference
94l PD 94tr PR/SPL 94b SI 95t SI 95bl SI 95r SI 96tl PD 96bc Clipart.com 96tr PD 97tl PD 97tr Wikimedia/Rama 97b PD 98tl PD 98bl PD 98c Wikimedia/Ballista 98b NOAA 99tl USDS/Susan Winchell-Sweeney, Laurie Rush 99c Lisa Purcell 99tr PD 99bl PR/Pascal Goetgheluck 99b NASA 100tl NOAA 100b USGS 100tr NASA 101tl NOAA 101r ESA/NASA/JPL/University of Arizona 101b PD 102tl PD 102l PD 102c PD 103 PD 104 CMGW/Ph. Bouysse 106tl SS/James Knopff 107 Pleum Chenaphun 108 Clipart.com 108l SS/Terry Alexander 108bc SS/Bateleur 108r SS/Adrian Hughes 109tl SS/Cecilia Lim H M 109tr SS/Rick Parsons 109br NOAA 109bl SS/Ugorenkov Aleksandr 110tl SS/Roman Krochuk 110r SS/Roman Krochuk 110bc NASA/JPL/Cal-Tech 111tl Rickley Hydrological Company 111r USGS Earthquake Center 111tr Clipart.com 111b SS/Rade Lukovic

Chapter 7: Go With The Flow
112 SPL/Chris Paola 113t SS/Koval 113b SS/Andrey Shchekalev 114tl Wikipedia/Alex Buirds 114l NWS 115tr USGS 115b NASA/USGS 116tl NRCS 116c SS/Dainis Derics 117t NRCS 117b NRCS/Bob Nichols 118tl SS/Aaron Kohr 118bl Wikipedia/Carlos Ponte 118r EPA 119 Wikipedia/Pollinator 120tl PD 120b Wikipedia/Daniel Ortmann 121l PD 121br SS/Wade H. Massie 122tl USGS 122r Howe Caverns, Inc. 122b Wikipedia/Hugo Soria 123t SS/Vova Pomortzeff 123r Freer Sackler Gallery, SI 124tl PD 124bl LoC 124tr PD 125tr NASA 125br PR/Raven

Chapter 8: Glacations Through Time
126 SS/Bryan Busovicki 127b SS/Lawrence Beck 127t PD 128tl LoC 128c PD 129tr Bob Kopp/Joe Kirschvink/Cal Tech Division of Geological and Planetary Sciences 129b SPL 130tl SS/Galyna Andrushko 130b SPL/Chris Butler 131t Wikipedia/Dschwen 131b Shuhai Xiao/Virginia Polytechnic Institute State University 131t IS/Vladimir Melnik 132c NASA 132bl Pleum Chenaphun 133 MF 134tl SS/Svetlana Privezentseva 134bl SS/David Lewis 134c Wikipedia/ Brendan Conway 135t USGS 136tl NASA 136l SPL/Munoz-Yague/Eurelos 136bl NOAA 137 NASA/Susan Twardy 138tl NASA 138b NASA 139t NASA 139b NASA/Craig Attebery

Chapter 9: Geological Catastrophes
140 SS/Bjartur Snorrason 141t NOAA 141b SS/Ian Bracegirdle 142tl Wikipedia 142c SIBL/NYPL/SPL 142bl SS/Bryan Busovicki 143tl USGS 143b USMC 144tl USDI 144l LoC 144br Clipart.com 145t SPL/Gary Hincks 145b PD 146tl Wikipedia/Kanoa Withington 146b University of Hawaii 147 PR/NASA 148tl IS/Gina Smith 148c SS/A.S. Zain 149tl PD 149bl NOAA 150tl Wikipedia/Ivelin Minkov 150b Wikipedia/Teri J. Pieper 151t USGS 152tl SS/ Marco Regalia 152bl PD 153t USGS 153b Wikipedia/Altidude 154tl NASA/Virgil L. Sharpton 154bl PD 154r PR/Detlev van Ravensway 155tr ESA 155br NASA

Chapter 10: A History of Geologic Thought
156 PR 157t SS/Sean Gladwell 157b LHL 158tl clipart.com 158c SI 158bl Wikipedia 159 SS/Vladimir Korostyshevskiy 160tl USGS 160c PD 161bl PD 161tl LHL 162tl SS/PMLD 162r LHL 162bl PD 163 SS/Stephen Aaron Rees 164tl PD 164b PD 165t Lisa Purcell 165b PD 166tl University of Washington 166b NASA/JSL 167tl AP Photo 167t SPL/D. Van Ravensway 168tl Pleum Chenaphun 168bl Ecolo.org 168tr Pleum Chenaphun 169tl Wikipedia/Javier Pedreira 168b SS/Elena Ray

Chapter 11: The Plate Tectonic Revolution
170 NASA/GSFC/PR 171tl NOAA/James McVey 171b Wikipedia/Chris73 172tl Pleum Chenaphun 172r SPL 172 Pleum Chenaphun 173 SPL/ Martin Land 174tl NOAA 174bl Pleum Chenaphun 174r NOAA 175 NOAA 176tl NASA 177b SPL/Gary Hincks 177 NASA 178tl SS 178tl SS/Bychkov Kirill Alexandrovich 178r SS/Dan Lee 178b NASA 179tr NASA 179b SS/Vladimir Korostyshevskiy 180tl NASA 180l Views of the World 180bl Wikipedia/JC Murphy 181tl PD 181r NASA 182tl SS/Dhoxax 182b SS/Rodolfo Arpia 183 SS/Bryan Busovicki

Chapter 12: Geology in the Field
184 PD 185t SS/Scott Rothstein 185b NASA 186tl SS/John Montgomery Brown 186l SPL/Sinclair Stammers 186br Brunton Inc. 187tr Pleum Chenaphun 187b Wikipedia 188tl USGS 188b USGS 189 NPS 190tl SS/Neo Edmund 190l NASA/EPA/John Holdzkom and Jim Szykman 191t USGS/Serkan Bozkurt 191b ESA 192tl NOAA 192b IS/FRONTIER Henri 193tr IS/Mike Morley 193b SS/Morozova Tatiana

Further Reading:
204 IO/photolibrary.com

At the Smithsonian:
206 SS/Vladimir Ivanov 207 PR/Martin Land 208 SI 209t Wikimedia/David Bjorgen 209b Peter Jones

Cover:
IO/DesignPics Inc. **Background:** IO/Hot Ideas